RULES OF THE GAME

ALSO BY NEIL STRAUSS

Emergency

The Game

The Long Hard Road out of Hell
WITH MARILYN MANSON

The Dirt
WITH MÖTLEY CRÜE

How to Make Love like a Porn Star
WITH JENNA JAMESON

Don't Try This at Home
WITH DAVE NAVARRO

How to Make Money like a Porn Star
WITH BERNARD CHANG

THE STYLELIFE CHALLENGE

THE ROUTINES COLLECTION

and

THE STYLE DIARIES

Neil Strauss

itbooks

A box set edition of *Rules of the Game*, containing "The Stylelife Challenge" and "The Style Diaries," was published in 2007 by HarperCollins Publishers.

HarperCollins books may be purchased for educational, business, or sales promotional use. For information please write: Special Markets Department, HarperCollins Publishers, 10 East 53rd Street, New York, NY 10022.

First It Books edition published 2009.

Illustrations by Bernard Chang
Designed by Jamie Putorti

Library of Congress Cataloging-in-Publication Data is available upon request.

ISBN 978-0-06-191169-9

13 DIX/RRD 10 9

READ ME

Are you that weak?

You're just going to allow yourself to get ordered around like that? You see the words "read me" at the top of a page, and you just follow the instructions like a robot?

I know what you're thinking: This is a book. You're supposed to read it.

But the book hasn't even started. Did you read the copyright information? Probably not. (If you did, welcome to the club. They make medicine for people like us.)

As long as you're here, though, I'd like to teach you your first lesson, which is to think for yourself. Not to just slavishly turn the pages of a book like a follower.

Because this isn't an ordinary book, meant to be read from front to back.

You have options.

If you're here to read, turn to *The Style Diaries*, which begin about two thirds of the way into the book, for stories compiled from my journals.

If you're here to learn, start with the first part of the book, *The Stylelife Challenge*, for a workout program for your social and dating skills. Then, whenever you're in the mood, dip into *The Style Diaries* for stories about the benefits and consequences of the knowledge you're accumulating.

If you need additional training wheels to help find just the right words to say, I've added *The Routines Collection* specifically for this new edition of the book. It's a compendium of word-for-word scripts of some of the most requested material I used while learning to put a permanent end to years of lonely nights.

And if you're just browsing through this book in a store for a quick tip on meeting someone new, tear out this page and fold it into a paper plane. Then throw it into the next aisle. You may do the same with all the following pages, until you meet someone special, most likely a security guard or police officer.

Finally, I'd like to take back what I said at the beginning of this section. You're not weak. I just wanted to get your attention. And bullying always seems to work best.

So perhaps it's time for your second lesson: The secret to life is not to take it personally. The things people do and say are not always about you. Usually they're about their own fears.

Thanks for reading . . .

CONTENTS

THE STYLELIFE CHALLENGE

To your mother and father. Feel free to blame them for everything that's wrong with you, but don't forget to give them credit for everything that's right.

CAUTION: DO NOT READ

THE TEMPTATION TO READ THIS BOOK COVER TO COVER
IN A FEW SITTINGS MAY BE STRONG.

THAT'S NORMALLY HOW BOOKS WORK.

NOT THIS ONE.

FOLLOW THE INSTRUCTIONS ONE DAY AT A TIME.
STUDY THE ATTACHED BRIEFINGS.
PERFORM THE FIELD MISSIONS.

AND DO NOT SKIP AHEAD.

MISSING A SINGLE LESSON OR EXPERIENCE WILL AFFECT
YOUR RESULTS, YOUR GAME, YOUR LIFE.

YOU'VE BEEN WARNED.

FEEDBACK FROM PAST PARTICIPANTS IN THE STYLELIFE CHALLENGE

"All I have to say about the Challenge is, 'Wow.' Before this month, I had never approached a woman or been on a date. I just had three dates in three days, and I have more numbers to follow up on."

—CHALLENGER NAME: DIABOLICAL

"This Challenge has been the most inspirational month in my entire life. I feel like I've achieved so much! Taking into account that all of this happened in just thirty days makes it unbelievable!!! Seriously, this was the only area in my life that kept me from being absolutely at peace or absolutely happy: women!"

—CHALLENGER NAME: TONY23

"Since the Challenge, I've heard I am a winner, amazing, perfect, one of a kind, her soul mate, and unbelievable! Thanks, Neil, for the Challenge!"

—CHALLENGER NAME: GODROCK73

"This is one of the best things I have ever participated in . . . It has been life-changing."

—CHALLENGER NAME: MAIDENMAN

"I've gotten more great responses from girls this week than I've ever had in my whole life. People I know are already telling me I'm different and charismatic now."

—CHALLENGER NAME: SAMX

"I already have a girlfriend, and I have no problems with girls. So why did I do the Challenge? Self-improvement. I must say, it has been an incredible boost, both to my self-confidence and to how others see me. I work as a waiter, and now customers are asking for me, girls are constantly approaching me, my tips are much bigger, and people want me to join their parties all the time. Everyone wants to be with me and in my inner circle, and everyone notices how good I feel about myself—and it wasn't like this before."

—CHALLENGER NAME: RACEHORSE

"She was French-kissing me and asking if she could see me when she gets back on Tuesday. I can't wait. She's not only beautiful but smart and kind. Were it not for Neil and the Stylelife Challenge, this never would have happened."

—CHALLENGER NAME: APOLLO

"Neil Strauss has given me a gift of life. I can't describe anything better. Just thanks."

—CHALLENGER NAME: LIZARD

"Waking up each morning is a treat since I started the Challenge. I have a certain excitement about me, like a child on Christmas morning getting ready to open presents, as I march upstairs to my office to see what's next for today. It's really an experience I'll never forget."

—CHALLENGER NAME: REIGN STORM

"This has been one of the most incredible experiences I have had in my entire life. Thank you so much for literally changing my whole life."

—CHALLENGER NAME: BOY

"This is one of the most rewarding experiences of my life! I am so far out of my comfort zone, words can't describe it! And I'm having more and more fun!"

—CHALLENGER NAME: GRINDER73

"I have done a lot of what I would consider to be intense things in my life, but in a way, this beats just about anything I've done, because it is literally changing my own perception of reality and what is in the realm of possibility for me. I would like to live the Stylelife Challenge every month."

—CHALLENGER NAME: LPIE

"Thank you, Neil . . . You'll be remembered for this forever. This is not just another book or a seminar. This is a really big deal!! One day I'll shake your hand and find a way to really thank you for changing my life!"

—CHALLENGER NAME: GRAND

"Neil, thanks a lot for this life-changing experience . . . Your efforts have had a great impact on my life. Not only will I use the information that I learned in my love life but also in all other aspects of my life."

—CHALLENGER NAME: BYRON

"To be honest, I never thought Neil was going to pull this off. I mean, successfully breaking down the seduction game in thirty steps isn't exactly the easiest thing to do. But Neil did an awesome job. Great material. Great people. Great results."

—CHALLENGER NAME: VELOS

"I have read plenty of dating books and seduction manuals. I think the material presented in the Stylelife Challenge is simply the best of the lot. Kudos to Neil for offering the best."

—CHALLENGER NAME: ALBINO

"I have to say, if you are serious about getting the dating aspect of your life handled, and you consciously don't do, or even read, Neil's Stylelife Challenge, then you need to really look at yourself and ask yourself what it is you truly want. Neil is giving us what no one in history has given."

—CHALLENGER NAME: BIG SEND

CONT

ENTS

INTRODUCTION

Why Are We Here?

I didn't want to write this book.

In fact, it's something I thought I'd never do.

I am as embarrassed to write this as you may be to pick it up. And that's fine. It means we're in this together.

Let me tell you why I'm embarrassed. Then I'll tell you why you're embarrassed. And then we'll agree to move on and recognize that we're here on the same page for a reason.

I spent my teens and most of my twenties lonely, desperate, and woefully inexperienced, sitting mutely on the sidelines while women obsessed over guys whose appeal boggled me.

At the lowest point in my dating career, after a two-year dry spell, I actually started surfing mail-order bride catalogs on the internet—Russian, Latin, Asian—bookmarking the pages of girls I thought I could learn to live with. I believed there was nowhere else to turn.

But then I had a reality-shattering experience—one of those moments that altered the course of my life. I discovered a secret society on the internet where men reputed to be the best pickup artists alive met to share tips, tales, and tactics learned in clubs, streets, and bedrooms around the world.

Emboldened by desperation, I disguised my identity, knocked on the door of that world, and it slowly opened. Inside, I dropped prostrate before the masters. I thought they would have the keys to release me from the prison of my own frustrations, fears, and insecurities.

They didn't have those keys. But I wouldn't trade the journey I took for anything. Because it taught me something I never would have realized on my own: that I actually had the keys the whole time. I just didn't know where to find them or how to use them.

When I wrote my account of those years, *The Game*, I thought it was my last word on the subject. I wanted to walk away gracefully. Even though I inadvertently became the top-ranked expert in the pickup community, I prefer to be a student of life, not a professor. I write not to teach but because I enjoy storytelling.

However, this book is not a story, at least not in the proper sense of the word. It is a how-to book. The story is not mine to write, but yours to live. The pages are turned not by plot, but by your own motivation.

The fitness and health industries offer thousands of programs designed to help you reach your physical goals. And there is an enormous and well-established self-help industry for women. The pages of *Cosmopolitan*, the characters on *Sex and the City*, and countless books, talk shows, and businesses exist almost solely to help cope with the challenges that come with being a woman in this world.

The landscape for men, however, is very different. Male sexuality is catered to everywhere in society—from the pages of *Maxim* to billboards selling the good life to the $97 billion porn industry. Everywhere they turn, men are shown images of women they are supposed to desire. Yet there is little advice of substance available to help them learn to attract these women, to figure out who they are, to help them improve their lifestyle and social skills. And considering that our social skills determine the course of our lives—our careers, our friends, our family, our children, our happiness—that's a big area to neglect.

So, even though I had no such intentions when I wrote *The Game,* I started doing a few things in my spare time to help the many guys who reached out to me after its publication, with emails, calls, and letters full of heart-wrenching stories. I coached frustrated teenagers, thirty-year-old virgins, recently divorced businessmen, even rock stars and billionaires. However, the more people I helped, the faster my inbox filled with requests from every corner of the world. Hundreds turned to thousands turned to tens of thousands turned to hundreds of thousands. And most of these guys were not assholes and creeps, but nice guys—the ones women always say they're looking for yet at the same time never seem attracted to.

So I decided to bite the bullet. You're now holding that bullet in your hands.

The Stylelife Challenge is a simple, easy-to-follow guide to approaching and attracting women of quality. Though it is designed for the hardest cases, it has also been proven to work for men who are already successful with women.

There is no method, system, or philosophy behind the Challenge. It is simply what works best and fastest. I have now spent five years gathering this knowledge, living it, and sharing it. I've tested the specific material in this book on over thirteen thousand men of varying ages, nationalities, and backgrounds.

The result: a monthlong workout program for your social, attraction, dating, and seduction skills.

I call it the Stylelife Challenge because it is my challenge to you: *Learn the game in thirty days.*

And I'm hesitant to do this, because just that last sentence alone sounds like I'm turning into one of those guys you see grinning from the covers of self-improvement books.

But if it helps you, then it's worth it. And in thirty days, we can both get on with the rest of our lives.

Now let's move on to your story.

HOW TO PLAY THE GAME

Throw out everything you know about dating.

If you're reading this book, it's because something in your life hasn't been working. And if something isn't working, there's only one way to fix it for good: Take it completely apart and rebuild it piece by piece. Only then can you make sure that every single component is functioning at its highest level, free of error, with the most up-to-date technology.

So if you're too intimidated to approach women you're attracted to, if you're a virgin, if you've never had a real girlfriend, if you're terminally shy, if you're recovering from a rough breakup or divorce, if you're suffering from a long dry spell, if you're tired of watching other guys have all the fun, if you want to attract higher-quality women, or if you're good with women but still not good enough, welcome to the Stylelife Challenge.

My challenge to you is simple: *Get a date in thirty days.*

Along the way, whatever your experience level may be, you'll receive the skills, tools, confidence, and knowledge to meet and attract almost any woman, any time you want.

I want you to master this part of your life. And to make sure you do, I'm going to hold your hand and walk you through every step along the way.

Why am I doing this? Because after reverse engineering my transformation from lonely to oversexed to just-right, as described in *The Game,* I developed a shortcut that compresses years of learning into a month. It has worked not just for me but for thousands of men—transforming their success with women as well as their success in a much bigger game: life.

Overview

- **The objective:** Get a date in thirty days or less.
- **Who can play:** Anyone seeking more success with women.
- **The cost:** The price of this book—and the willingness to try on some new behaviors and see if they fit.
- **The prize:** The company of quality women, the envy of your peers, the lifestyle you deserve.
- **How to play:** This book contains thirty days of exercises. Set aside at least an hour a day—the days don't actually need to be consecutive—to perform the suggested missions and read the supplementary material.

Guidelines

Your instructions are simple: Every morning, as soon as you wake up, read your missions for the day. They may be primers to study, questions to answer, self-improvement exercises to perform, or field exercises to get you out of the house and approaching women. They begin at a very basic level and grow more advanced as the Stylelife Challenge continues. Think of it as a fitness program for your social life.

If you want to get the most from the Challenge—so that your friends and family will instantly notice the new you—it's important that you complete all of the missions in the order they are presented. Do not read ahead. Some exercises may seem basic; others may seem out of character for you. But each new exercise builds upon the last, so stick with it.

Several missions will require you to read certain guides and articles. These can be found in the supplementary briefing immediately following the breakdown of the day's tasks. Make sure that you read each briefing before proceeding to any corresponding field assignments.

The only other material you need will be a pen and paper—although access to a mirror, a computer with an internet connection, and some way to record your voice will be useful for a few assignments. You may also want to keep a journal.

You will not need any money to compete, but you will need a little time each day to do a few small things that can change your life in the long run.

None of these assignments requires much more than an hour, so even if you're working three jobs, you should still be able to do them all. In a pinch, you can always save time by cutting back on all that energy wasted desiring women from afar (in men's magazines, on TV shows, in the street, on the internet) and instead learning what it takes to have them in your life.

Though the Challenge is designed to be completed alone, if you're the type of person who's motivated by communicating with others on the same path, optional discussion boards are available at www.stylelife.com/challenge. You can post all questions, adventures, sticking points, and successes there. My trained coaches, your fellow Challengers, and I will be there to help you. In addition, you'll find video and audio examples demonstrating some of these exercises and approaches. Note that all the additional tools provided to supplement this book are free.

How to Win the Game

You win when, at any point between Day 1 and Day 30, you get a date.

A date is defined as a planned second encounter with a woman you have just met.

For example, if you approach a woman at a bar, exchange phone numbers, and meet her for coffee two days later, that is a date.

If you talk to a woman at the mall and arrange to meet that night at a bar, and she shows up specifically to see you, that is a date. Even if you don't exchange phone numbers.

Basically, any scenario where you approach a woman and she agrees to see you at a later date or time—and shows up—constitutes a date.

Once you get a date, feel free to put your name in the winner's circle at www.stylelife.com/challenge and share your story. If you win before the thirty days are up, feel free to continue the Challenge and carry out the daily missions for the remainder of the month. They'll only further enhance your confidence and game.

When you're ready to receive your first mission, turn the page and begin the Stylelife Challenge.

Enjoy, and play fair.

THE STYLELIFE CHALLENGE

DAILY MISSIONS

DAY

MISSION 1: Evaluate Yourself

Fitness programs require you to weigh in on the first day. Financial plans ask for a list of your assets and debts. So to revamp your social life, you'll need to make a social assessment of yourself.

Your first mission is to write answers to the following questions. Don't worry about what anyone else will think of your answers. Your goal is to be as honest with yourself as possible.

1. Write one or two sentences describing how you believe other people currently perceive you.

   ____

   ____

   ____

   ____

2. Write one or two sentences describing how you'd like to be perceived by others.

   ____

   ____

   ____

   ____

3. List three of your behaviors or characteristics you would like to change.

   ____

   ____

   ____

   ____

4. List three new behaviors or characteristics you would like to adopt.

____

____

____

____

MISSION 2: Read and Destroy

Before moving on to your first field assignment, it's necessary to eliminate any self-sabotaging beliefs that you may have about interacting with women. Your next task is to read the manifesto titled "The Chains That Bind," included at the end of today's assignments in the Day 1 Briefing.

MISSION 3: Operation Small Talk

Your first field assignment: Make small talk with five strangers today.

It doesn't matter whether they're male or female, young or old, friendly or unfriendly. The stranger can be a businessman in the street, an old lady in the supermarket line, a hostess at a restaurant, or a homeless person.

The goal is simply to start a conversation, with no intent other than filling in the silence with a question or pleasantry. The conversation doesn't have to progress beyond a comment and a response.

If idle chatter doesn't come naturally to you, scan news headlines before you leave the house. Small-talk topics include:

- Weather: "It's beautiful out today. Too bad we're stuck inside."
- Sports: "Did you catch the ________ game last night? I couldn't believe it."
- Current events: "Did you hear that ________? What are they going to think of next?"
- Entertainment: "Have you seen the new ________ movie yet? I wonder if it's any good."

Remember: The answer doesn't matter. Whether you receive a long story or a cursory grunt in response, you've completed the mission simply by opening your mouth and speaking to a stranger.

DAY 1 BRIEFING
THE CHAINS THAT BIND

When it came to meeting women, my biggest enemy was me.

I used to look at myself—five foot six, scrawny, bald, and big nosed—and think there was no way I could compete with all the tall, good-looking guys out there. I was so unhappy that I considered plastic surgery.

But once I started approaching women in streets, bars, clubs, and cafés, I discovered that looks don't matter nearly as much as I'd thought. As long as I was well groomed, all I needed in order to attract just about anyone I wanted was the right personality.

Although it's a dubious achievement to be named in the media as the best pickup artist in the world, one thing it taught me was that I didn't need to change the way I looked. I was doing just fine. In fact, I usually had it easier than big, muscular, square-jawed male models because I was much less threatening and intimidating. I could come in under the radar. In the end, then, my problem wasn't my looks, but my limiting beliefs about my looks.

A limiting belief is something that you believe about yourself, other people, or the world—and although it isn't actually true, the fact that you *think* it is holds you back from experience and success. Any time you tell yourself you "can't" do something that's within the realm of human possibility—that's a limiting belief.

Dispelling limiting beliefs is very easy: Just ask yourself, "Was there ever a time when . . ." and insert your limiting belief. For example, if you believe that you get uncomfortable around beautiful women, ask yourself, "Was there ever a time when I was comfortable around a beautiful woman?" Name just one time, and you've disproved your limiting belief.

Nearly everyone is held back by some limiting belief, whether he's conscious of it or not. So before I send you running around the streets talking to strangers, let's clear the air and dispel a few of the most common limiting beliefs about dating.

LIMITING BELIEF: If I talk to her, she'll ignore me—or, even worse, say something mean that will embarrass me.

REALITY: Here's something that may surprise you: The harder it is for you to approach women, the less likely it is that you'll be rudely rejected.

Why is that? Because most people have been raised to be courteous and polite, unless they feel threatened—and a shy guy isn't too likely to intimidate anyone. The worst thing that's likely to happen is the woman will politely say she's having a private conversation, or simply excuse herself to go to the bathroom. Playing negative what-if scenarios in your head is detrimental to your emotional health. Instead, get out of the house and start approaching women, and you'll discover that most of the things you imagine going wrong will never happen.

LIMITING BELIEF: People are looking at me, judging me, or making fun of me.

REALITY: This is half right. People may notice you, but they're not necessarily judging you—most of them are too busy worrying about what other people are thinking of them. Once you realize that most people are just like you—and that they're actually seeking your approval—you'll start to become socially fearless.

Besides, most bystanders who see you approach a girl or a group assume that you know the people. So act like you do. Not only will it ease your worries about what everyone else is thinking, but it'll also make your approach more effective.

LIMITING BELIEF: Women aren't attracted to nice guys. They like jerks.

REALITY: This is one of the oldest myths about dating. And, fortunately, it's inaccurate. The dating dichotomy isn't actually between nice guys and mean guys, or good boys and bad boys. It's between weak guys and strong guys. Women are drawn to men who demonstrate strength—not necessarily physical strength, but the ability to make them feel safe. So if you're a nice guy, you can still be nice. But you must also be strong.

However, make sure you know what nice means. Most guys who define themselves as "too nice" only behave nicely because they want everybody to like them and don't want anyone to think badly of them. So, if this is you, get off your nice high horse. Don't mistake being fearful and weak-minded for being nice.

LIMITING BELIEF: I'm not good-looking, rich, or famous enough to be with a beautiful woman.

REALITY: There are plenty of rock stars and multimillionaires who have the exact same problems with women that you do. I know because I've coached many of them. And, in the process, I learned that money, looks, and fame—while they certainly make things much easier—aren't actually necessary. Fortunately for men, the way we look doesn't matter nearly as much as how we present ourselves. And this requires only good grooming, and clothing that conveys an attractive identity. When it comes to wealth and fame, simply displaying the desire and ability to achieve them can be just as powerful. Like talent scouts, many women are attracted to men with goals and potential. And in the next ten days, we'll be sharpening your appearance, goals, and perceived potential.

LIMITING BELIEF: There's this one girl . . .

REALITY: There are many incredible women in this world. If you're hung up on one particular girl you just can't get out of your mind—and she hasn't given you any sense that she shares the feelings—then recognize that's not love you're feeling, but obsession. And that obsession is likely to scare her away. The best thing you can do for yourself and for her is to go out and interact with as many women as possible, until you realize that there are plenty of people out there for you—some of whom are capable of recognizing your worth and reciprocating your feelings.

LIMITING BELIEF: Some guys are born with the ability to charm women. Other guys just don't have it and never will.

REALITY: Fortunately, there is a third type of guy: one who can learn it. That's me. And once you understand how attraction works and have a few successful approaches under your belt, it'll be you too. Any problems you may currently be having aren't the result of who you are but of what you're doing and how you're presenting yourself. Those problems can be fixed easily with the right knowledge and a little practice. If you stick with the program after the Challenge, you'll even start doing better than the so-called naturals you once envied.

LIMITING BELIEF: All I have to do is "be myself," and eventually I'll meet the right woman who likes me for me.

REALITY: This works only if you know exactly who you are, what your strengths are, and how to convey them successfully. Most often, this statement is used as an excuse not to improve. What most of us present to the world isn't necessarily our true self: It's a combination of years of bad habits and fear-based behavior. Our real self lies buried underneath all the insecurities and inhibitions. So rather than just being yourself, focus on discovering and permanently bringing to the surface your best self.

LIMITING BELIEF: To figure out what women want, just ask them.

REALITY: This may be true sometimes, but not as often as many people think. It wasn't until I started trying behaviors that seemed counterintuitive that I discovered a key principle of the game: What women want isn't necessarily what they respond to. Furthermore, what women *say* they want may be what they want in a relationship, but it isn't always what attracts them during the courtship period. That said, most women will give you the information you need to attract them, but it's usually found between the lines.

LIMITING BELIEF: If I approach a woman, she'll know I'm hitting on her and think I'm lame.

REALITY: This is only partially true—women think this only when men approach them *badly*. This includes men who make them uncomfortable, creep them out, or seem to have an agenda. The biggest mistake a man can make with a woman is hitting on her before she's attracted to him. And though this describes the so-called technique of most men, it's a mistake you'll avoid if you follow your daily missions. Few women will resent meeting someone who is warm, funny, sincere, interesting, engaging, makes them feel comfortable, and isn't going to stick around talking their ear off.

LIMITING BELIEF: Women don't like sex as much as men do. They're mostly interested in having a relationship.

REALITY: If you believe that, you haven't spent enough time around women. Here are a few facts that may help dispel that belief: It's women, rarely men, who have an organ solely made for sexual pleasure: the clitoris, which has twice as many nerve endings as a man's entire penis. And it's women, not men, whose orgasms can last minutes or longer. Most men have just one orgasm and then lose their arousal; most women can have orgasm after orgasm and many different types: clitoral, vaginal, blended, full-body, and psycholagnic (look it up).

In short, good sex is even better for women than it is for us. So doesn't it make sense that they want it more?

DAY

MISSION 1: Set Your Goals

Congratulations! You survived Day 1.

Whether you already know your life goals or you just need a little prodding, today's first exercise will help you set your intent and program your mind for success.

To quote J. C. Penney, founder of the department store chain, "Give me a stock clerk with a goal, and I'll give you a man who will make history. Give me a man with no goals, and I'll give you a stock clerk."

Your mission is to read the following questions, think about them carefully, and write your personal mission statement. Be as specific and ambitious as possible. (Examples of accomplishments include starting a band, buying a house, getting in shape, launching a business, becoming president.)

1. What three accomplishments would you like to achieve to make you happier?

   ____

   ____

   ____

   ____

2. What are the reasons these accomplishments will make you happier?

   ____

   ____

   ____

   ____

3. What is your personal mission?

I will become ______________________ (maximum four words)
MY ROLE

who will ______________________ (maximum four words)
MY CLAIM TO FAME

within ______________ days/weeks/years.
NUMBER

4. List three specific results that will let you know that you've accomplished your mission. (For example, "I will have earned $200,000," "I will have lost thirty pounds," or "I will have won five Academy Awards.")

1. I will have ______________ ________ ________.
ACTION VERB NUMBER ASPECT

2. I will have ______________ ________ ________.
ACTION VERB NUMBER ASPECT

3. I will have ______________ ________ ________.
ACTION VERB NUMBER ASPECT

5. Why are you now fully committed to pursuing your personal mission?

Because if I *don't* pursue it *now,* I will continue to suffer over the next years and

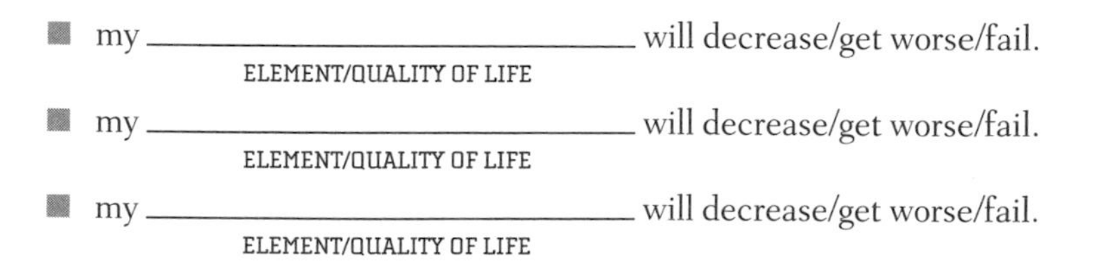

- my ______________________ will decrease/get worse/fail.
ELEMENT/QUALITY OF LIFE
- my ______________________ will decrease/get worse/fail.
ELEMENT/QUALITY OF LIFE
- my ______________________ will decrease/get worse/fail.
ELEMENT/QUALITY OF LIFE

But if I *do* pursue it *now,* I will enjoy the next years and

- my ______________________ will increase/improve/come true.
ELEMENT/QUALITY OF LIFE

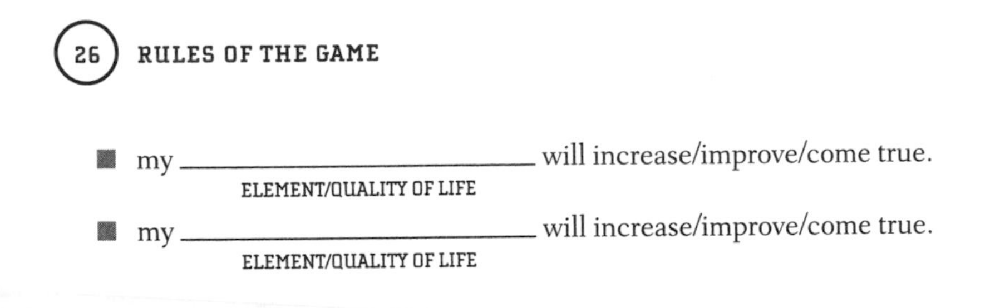

- my ________________ will increase/improve/come true.
  ELEMENT/QUALITY OF LIFE
- my ________________ will increase/improve/come true.
  ELEMENT/QUALITY OF LIFE

MISSION 2: Look into Your Eyes (Optional)

There's another step you can take to reinforce your personal mission statement and strengthen your subconscious intent: self-hypnosis. I've commissioned a charismatic mind-shaping exercise specifically for the Challenge, which I've made available for you online at www.stylelife.com/challenge.

After you download it, find a comfortable place free of distraction. Dim the lights, take off your shoes, and sit or lie down. Relax. Then put on headphones, play the audio, and take the journey.

Make sure you listen to the entire recording without interruption. It's more important to *feel* this experience than to see it. Try to listen to the recording every other day during the Challenge: The more you repeat it, the better the result.

MISSION 3: Look into Their Eyes

Your field assignment today is to go out and make small talk with five more strangers.

But, this time, there's one more thing you need to do: make eye contact with each person. Record his or her eye color in the space below:

1. ________________
2. ________________
3. ________________
4. ________________
5. ________________

In the first small-talk exercise, the purpose was to develop the ability to talk to anyone without fear. Meeting people eye to eye (being careful not to stare) will not only increase the likelihood of a response, it'll help you connect with them on a more personal level.

If you'd like to develop this crucial but subtle skill further, here's an extra-credit exercise: Try to hail a cab, get a bartender's attention, or call a waiter to your table without speaking or gesturing—instead, use nothing but eye contact.

MISSION 4: A Hint for Tomorrow

Be sure to read tomorrow's assignment the moment you wake up—before you shower, shave, or check your email.

DAY

MISSION 1: Adopt the Caveman Hygiene Method

This next mission may make you a little uncomfortable. And that's a good thing. The reason will be made clear tomorrow. But for now:

Do not shower today.

Do not shave today.

Chances are, no one will notice—most people are too busy worrying about how they look. If they do, tell them you're trying to win a bet or participating in a highly compensated study for the deodorant industry.

MISSION 2: Speak with Confidence

When I was learning the game, I had trouble meeting new people because I talked too fast, too softly, and swallowed my words. In a loud club, it made meeting women practically impossible. So I went to a vocal coach named Arthur Joseph.

"Your voice is your identity," he teaches. "It can tell people everything about who you are, how you feel about yourself, and what you believe in."

So today we're going to work on your voice.

There are five common speech mistakes people make. These errors are outlined, along with an exercise for each, in your Day 3 Briefing.

Your task is to read the article and do at least three of the exercises, even if you don't think you need to. You may be surprised.

MISSION 3: Find Mr. Moviefone

For today's field mission, stay home. You're going to use only your voice.

Your task is to dial a local number randomly on your telephone. When someone answers, try to get him or her to recommend a good movie. That's all.

The point isn't just to talk to more strangers. It's to learn how to change the course of an interaction without making the other person feel uncomfortable.

This skill will help you take control of conversations in real life and direct them toward the outcome you want.

A few hints:

Rather than just dialing random strings of seven-digit numbers, look through a residential telephone book and select numbers at random. Or use the first three digits in your own number and make up the last four digits.

Here's a sample script I used when doing the Challenge myself:

"Hi, is Katie there? No? Well, maybe I can quickly ask you this instead." Don't pause here and give the person an opportunity to say no. "I want to see a movie tonight. And I was wondering, have you seen any good movies lately that you'd recommend?"

Here's another script that worked:

"Hello? Is this Moviefone? No? Well, would you mind quickly recommending a movie to watch tonight? Have you seen anything good lately?"

If the person you're speaking to hesitates or asks if this is a joke, reassure him or her by saying that you're serious. One magic word you can use is *because*. Providing a reason, no matter how illogical (such as "No, I'm serious, because I'm in a rush"), psychologically influences people to accept an unexpected behavior.

Once you've received a movie recommendation from three separate people, consider today's mission successfully completed.

MISSION 4: Hypno Time (Optional)

Listen to yesterday's charismatic mind-shaping exercise again. Understand and begin to integrate your new attributes and self-image.

DAY 3 BRIEFING
VOCAL TRAINING

With the help of several vocal coaches, I've put together five exercises designed to eliminate weakness in your speech and bring out your most full, powerful, and commanding voice.

Before beginning this activity, you'll need:

- A mirror, preferably full-length;
- An audio recorder, or a computer with a microphone;
- An open area where you can be loud.

The Basics

There are two factors that make all the difference between a good orator and a bad one: breath and posture.

Breathing deeply before you speak fills your lungs with air, allowing you to give full power to your words. To ensure that you're doing this correctly, take a deep breath. If your chest expands, your breathing is too shallow.

Try it again until your diaphragm—the sheet of muscle beneath your rib cage—expands. To check this, place your hand on your stomach to make sure it rises with each inhalation.

Bad posture can restrict your diaphragm and breathing, effectively neutering your vocal power. Whenever you speak, make sure that your upper body is straight and aligned. If necessary, use the technique of imagining a string running from the bottom of your spine to the top of your head and then pulling it taut. But don't get too tense; make sure you're relaxed comfortably into the frame of your body. If this seems unnatural, don't worry: Tomorrow we'll examine your posture in detail.

PROBLEM: Low or Soft Voice

SOLUTION: Find a large, open space indoors or outdoors. Bring an audio recorder, a trusted friend, or both.

Take three large steps away from your audio recorder or friend.

Take a deep breath from your diaphragm. Hold it, then slowly exhale.

Take two more deep breaths. Then inhale one more time, and as you exhale, say, using your everyday voice, "I can say this without shouting and still be heard."

Now go back and listen to your voice on the recording, or ask your friend how you sounded.

Return to the same position and recite the same line. This time, instead of speaking to your friend or the recorder, aim your voice at a spot six to ten feet above. Imagine your voice is a football, traveling a wide arc to make a field goal. Afterward, check the results for improvement.

Take three more large steps away and repeat the same sentence: "I can say this without shouting and still be heard." Try to increase the volume of your voice without screaming or changing the tone.

Take another three steps away. Remember to send your voice in a high arc, past the listener. Afterward, listen to your recording (or your friend's reaction) and critique your vocal projection. See how far away you can stand and still be heard clearly without shouting. Practice this until you're comfortable talking at loud volumes without changing the tone of your voice. You'll notice that, in the process, you'll begin to speak more clearly as well.

If you've been a quiet talker all your life, chances are that the volume of your voice in your head isn't the volume at which other people hear you. So if you normally talk at a 5, from now on take it up to a 7. Don't worry about speaking too loudly. It's much more likely that your friends will start complimenting you on how clearly you've started communicating.

PROBLEM: Fast Speech

SOLUTION: Speaking too rapidly is one of the most common and crippling vocal mistakes. Not only does it make you difficult to understand, but it gives others the impression that you're nervous, you're not confident, and what you have to say is unimportant.

A calm, slow voice commands authority.

For this exercise, sit up straight in front of your audio recorder or computer microphone. Take a deep breath. Now say without slowing down the following sentence—all in one breath: "I will no longer speak too quickly and cram all my words together in one breath because I have lots of thoughts in my head and I am trying to get them all out and I am afraid that if I pause, people will stop listening."

Listen to the recording. Most likely, cramming a run-on sentence into one breath worsened your enunciation and caused you to swallow some words.

Now inhale and say the same line. But this time, make the pace exaggeratedly slow and deliberate; leave excruciatingly long pauses between phrases; pronounce each word carefully; and take a breath more often than you feel you need to. Then listen to the recording.

Repeat this exercise five to ten times, gradually increasing the pace, normalizing your breathing, and shortening the pauses between words while making sure you're still speaking slowly and pronouncing each word fully. This is going to feel unnatural at first, but stick with it until you find a comfortable and clear speaking pace that captures the attention of others.

Repeat the run-on sentence several more times in front of a mirror until you get used to your new speaking pace.

After you've mastered this exercise on your own, your voice may well speed up again in social situations. So make sure you monitor yourself, and take a breath and slow down as soon as you catch yourself speed talking.

Just like turning up the volume on your voice, it may take a while for your inner ear to get used to this change. You may think you're boring others, but you're not. Fast speakers often discover that, even when they've slowed down to what seems like an interminable crawl, they're still talking faster than everyone else in the room.

PROBLEM: Brain Farts

SOLUTION: Brain farts, or pausers, are the enemy of confidence.

Whether or not you know what a brain fart is, try this exercise before reading any further: Record yourself speaking with a friend. Either take an audio recorder with you when you leave the house, or record your end of the conversation next time you're on the phone.

Play back the recording and carefully transcribe the first few sentences. Make sure you write down every single word you say. Don't leave out anything.

Now take a look at what you've written. Do you notice the words *um* or *uh* anywhere? How about "you know," "like," or "whatever"? These are known as pausers, or brain farts.

We've learned to use these meaningless utterances for several reasons: as placeholders, to make sure we don't lose anyone's attention while we're thinking of what to say next, and as a sonar system, to make sure the other person understands or agrees with what we're saying.

But do you know what message these pausers actually send to others? Insecurity.

Pausing for a moment won't cause you to lose someone's attention. Always speak as if you're making complete sense—even when you don't think you are. The fact is, the way you communicate makes more of an impression than what you say.

Now listen to ten minutes of the conversation you recorded. Write down every pauser you say, then read them out loud (unless the sheet is blank, in which case you should apply for work as a newscaster immediately). Repeat them until they're imprinted in your mind so that you'll be conscious of them during future conversations. From now on, slow down and consciously choose each word when speaking.

The secret to eliminating pausers—and to breaking most other bad habits—is to become self-correcting. In other words, listen to yourself when you speak. If you notice a brain fart, stop, correct yourself, and repeat the sentence without the pauser. It may also help to carry your list of pausers with you, as a reminder to monitor your speech for these small signifiers of insecurity.

PROBLEM: Monotone Voice

SOLUTION: If you drone like an old geography teacher when you speak; if your friends close their eyes when you tell a story; if your colleagues tune out halfway through your presentations, you just may have a monotone voice.

Here's an excerpt from a children's short story. Read it out loud into your audio recorder now:

> *Leopold Elfin had a problem: His nose whistled. He couldn't help it. Every time he breathed through his nose, out came a note. Not the quiet hiss that occasionally issues from the hoary nostrils of men three times his age, but a loud, shrill shriek like a crossing guard blowing for traffic to stop. Leopold was well aware of this problem, but he'd never been to see a doctor, figuring it was more a matter of anatomy than medicine. Maybe it was his*

pinched septum, his narrow oval nostrils, or the crook at the bridge of his nose that was responsible for his one social impropriety.

Now play back the recording. If possible, listen to it with a friend or family member to get a more objective opinion.

Do you have a dynamic storytelling voice, the kind that sucks listeners into the world you're describing? Or do you have a monotone voice, the kind that listeners tend to tune out?

If it's the latter, then turn on the television. Find a male host, comedian, or other broadcaster with a dynamic voice that you like. Listen to him speak. Pay attention to every detail and nuance that make his voice compelling. Notice how he is present in the material, how his voice rings with energy, warmth, and immediacy.

Next, try repeating what he says, using exactly his words, tone, and style.

When you feel you're able to convey a few of his engaging qualities, go back to the story excerpt. Read it again into the recorder, using the techniques you just learned. Experiment with changing the volume, pitch, speed, timbre, rhythm, and flow of your voice as you read. Try emphasizing different words; creating pauses where they don't normally belong; shortening or elongating words; and speaking in different voices and accents. Read the excerpt several times, and don't be afraid to get silly if it helps you break through your limitations.

When you're finished, read the paragraph once more. This time, imagine you're recording a book on tape for children. Compare this new version to your original version—and discover the great storyteller lurking inside you.

PROBLEM: Statements that Sound like Questions

SOLUTION: Sit down, pull out your trusty recorder, and place it in front of you.

For your final vocal exercise, imagine that the audio recorder is your friend. And this friend of yours doesn't like fish. Your goal is to convince him to try sushi with you tonight.

When you're finished, play back the recording. Listen carefully. Does your voice rise in pitch at the end of any declarative sentences?

If it does, you'll notice that your statements sound like questions. And that makes you seem unsure of yourself.

Persuasive speakers end their sentences—and their argument—conclusively.

If your statements end in a higher pitch than they started, record the same speech again. This time, be firm. Instead of asking questions that beg for affirmation, make definitive statements that demonstrate your conviction. And make sure that the speech itself doesn't trail off into extraneous blather and repetition but instead comes to a definite and powerful conclusion. Sound like you know what you're talking about and believe every word you say. Even if you don't happen to like sushi.

When you have this mastered, you're done.

Congratulations.

However, just because you were able to identify and correct these five major vocal mistakes today doesn't mean the problem is solved for good. Revisit these exercises twice a week. And whenever you're in conversation, monitor your posture, breathing, and speech. If you catch yourself backsliding, correct yourself immediately. Before long, you'll not only have women hanging on your every word, you'll have your own radio talk show.

DAY

MISSION 1: Hit the Showers

As soon as you wake up, put on your favorite upbeat music and play it loud. Shower, shampoo, and soap thoroughly. Wash twice if you want. And . . . don't masturbate today, if you're prone to doing so.

Put something scented on your body: moisturizer, talcum powder, or a light spritz of cologne. Gargle with mouthwash. Whatever makes you feel and smell good.

Then shave your face clean (preserving any preexisting mustache, beard, or goatee). Make sure you shave or tweeze any stray places where you sprout hair—your ears, nostrils, the back of your neck.

Put on clean, well-fitting clothing. You should feel like a million dollars.

Now look at yourself in the mirror and read the following to yourself:

"You are amazing. People love you and respect you. You radiate charisma, charm, and grace. You stand out from everyone around you. Talking to you is a privilege. And you deserve the best the world has to offer. It's all there out there, waiting for you."

Read it as many times as it takes—say it out loud if you have to—until you truly feel and embody it.

Now hold on to that feeling . . .

MISSION 2: Ask an Expert

What you experienced in the previous mission is a simple ritual that helps many men enter a state of increased confidence, positivity, and unassailability. Take a moment to develop your own ritual to pump yourself up before going out to meet women. It may involve exercising, cleaning, repeating affirmations, reading something inspirational, replaying previous successes in your mind, blasting your favorite music, singing, showering, dancing, calling someone who makes you laugh, or any combination of the above.

This is the first day you're going to meet women you could possibly date. You should make these approaches at the earliest possible opportunity after leaving the house clean, well shaven, and feeling good about yourself.

Your mission: Ask three women to recommend a cool local clothing store that carries menswear. Your mission is complete once you've approached three women *and* received one clothing store recommendation. (In other words, if you approach three women and you get a clothing store suggestion, you're done. If you approach three women and you don't get a clothing store recommendation, keep asking until you do.)

When you get a recommendation, write down the name of the store, and the location if she knows it. Make sure you keep the name and location handy.

Here are a few tips:

- Approach women who seem like they live in town and have a cool sense of style.
- If you're talking to people in the street, don't approach them from behind, which can be startling. Either approach them from the front, or walk ahead of them and turn your head back over your shoulder as you keep walking. They'll feel even more comfortable if you increase the distance as you walk, as if you have somewhere to be. You may also approach in cafés, shopping malls, or wherever you're comfortable.
- Be aware that only about 1 in 3 women will be able to think of a store right away. Some people go blank when put on the spot.

As soon as she answers, even if it's just to say "I don't know," you've made your approach. Tell her "Thanks for your help" (or "Thanks anyway" if she doesn't have any ideas) and leave if you want. Or continue the interaction. The choice is yours.

Good luck.

MISSION 3: Stand Up Straight

Before you even open your mouth, a woman has formed an initial impression of you. And that impression is based largely on your body language. Today you're going to learn to carry yourself with confidence through a simple posture exercise known as the wall stance.

Stand with your back against a wall. Make sure your heels, butt, and shoulders are touching the wall. In addition, the back of your head just above the level of your chin should be against the wall.

Remain in this position for a minute. Reach behind your back and check to make sure there isn't too much space between your lower back and the wall. If there is, tighten your abdomen to bring the small of your back closer to the wall.

Now move away from the wall, and walk around the room for a minute without changing your posture. Commit the position and alignment of your body to memory.

Repeat this exercise one more time today and, if possible, once a day throughout the Challenge. From now on, check your posture on a regular basis, and bring yourself into alignment if you catch yourself slouching.

Because posture is key not just to your confidence and appearance but also to your health, I've prepared an extra-credit video tutorial for you online at www.stylelife.com/challenge. It provides the basics on Alexander Technique, a school of movement that improves not just the way you stand, walk, and sit but also the way you speak and feel about yourself.

DAY

MISSION 1: Here Comes the Groom

Today is grooming day, and the focus is your appearance.

When men discuss attraction skills, they often act as if looks are the only variable out of their control, perhaps because they feel that appearance is genetic. Not true.

Just as any girl can slim down, get breast implants, and dye her hair blonde to turn heads, any guy can become good looking. In the same way you can learn openers, routines, and confidence, you can learn looks. No matter how you're perceived right now, if you're willing to make a few changes, you can be considered good looking.

I've taken worst-case scenarios—fat, balding, acne-plagued guys in Coke bottle glasses—and through the miracles of tanning, contacts, head shaving, dermatology, health clubs, and menswear, turned them into cool, good-looking men who exude confidence and power.

Now it's your turn.

Your assignment is to read the grooming checklist in your Day 5 Briefing. Then perform at least one task on the checklist. Not all the suggestions will apply to you, so choose one from the area in which you're most deficient.

If you have a trustworthy female friend, ask her: "If you had to pick one thing to change about the way I groom myself, what would it be?" Let her know you'd sincerely appreciate an honest, constructive answer—and make sure you don't take it personally when she gives you one.

MISSION 2: Make a Change

The first step to better looks is better grooming. The second is committing to the right style.

Ideally, you want your style and clothing to convey that you belong to one of three segments of society: the same niche, group, or tribe that the woman you're interested in belongs to; a tribe she wants to belong to; or a tribe she wants to visit. For example, men in dirty oversized undershirts and ill-fitting khaki shorts belong to few women's tribes, while pierced, tattooed rock stars belong to a tribe that most women at least want to visit.

Thus, your mission today is to get a free style consultation.

Do this by examining the results of yesterday's field exercise and selecting the clothing store that received the highest recommendation. If possible, avoid large chains. Choose a small independent store instead.

Go to the clothing store—preferably when it's least likely to be crowded—and speak to the saleswoman who seems the most helpful. Tell her you want to change your style, and ask her to put together a complete outfit for you. If she wants you to be more specific, tell her you're going to a high-profile fashion show, art opening, movie premiere, trendy club, or whatever imaginary event best suits the new you.

Change into the new outfit and observe yourself in the mirror. Though the style of the clothing matters, a perfect fit is more important.

If you truly detest the clothing, tell her why and ask her to put together another outfit. If the saleswoman isn't helpful or pushes you too hard to buy, go to another store.

If you like the outfit and can afford it, buy it. When you get home, make sure you take care of it by hanging it in a closet and dry-cleaning it when it's dirty.

If the clothes are beyond your means, remember the brands, sizes, and styles, so you can either buy them in the future, find equivalent items at a used-clothing store, or order them cheaper online.

If you choose to buy the outfit, ask the saleswoman where you can find a nice pair of shoes. At the shoe store, show an employee the outfit and ask for sharp shoes that match it.

MISSION 3: Brush Up

Choose one of the following to experience again: the mind-shaping audio, vocal exercises, or posture wall stance and video. Try to review at least one of these fundamentals every day during the Challenge.

MISSION 4: Lay Out Your School Clothes

If you bought any new clothing or accessories today, be prepared to wear them tomorrow.

DAY 5 BRIEFING
GROOMING CHECKLIST

Choose at least one of the items on this list and make the suggested change. Not all of these tips will apply to everyone. Some are overly remedial; others are extremely meticulous. A few of the tips you'll be able to implement in just a few minutes at no cost; others may take time or money. Avoid the tasks you're most comfortable with. It's the changes you're uncomfortable with that will lead to the most improvement.

- **Change your hairstyle.** Look through music and men's fashion magazines, find the haircut you'd most like to have, and make an appointment at the best beauty salon in town. Bring the photo with you. Make sure you ask your hairstylist to recommend any hair product necessary to maintain your new look.
- **Ditch the glasses.** Get contacts or laser surgery. If your glasses complement your style, consider getting cool designer frames.
- **Get tan.** The quickest and easiest way to do this is to get a spray-tan at a tanning salon. Make sure they use a relatively realistic-looking brand, like Mystic Tan.
- **Get a manicure and pedicure.** Go to any nail salon. It isn't necessary to get a colored polish; just ask to get your nails buffed or request a clear top coat. Not only does this convey good grooming, but it will help you understand that the reason a woman pays attention

to the small details on you is that she pays attention to those details on herself.

- **Remove excess hair.** Get tweezers or a nose-hair trimmer, and remove any hair in your nostrils, between your eyebrows, in your ears, and on the back of your neck. If you're particularly hirsute elsewhere, trim it, shave it, or pluck it.
- **Examine yourself closely in a mirror.** If possible, buy a magnifying mirror. Remove any visible ear wax with a Q-tip; tweeze any stray hairs; clip and clean your fingernails and toenails; and look for oily skin, dry skin, bags under your eyes, or other problem areas that require the use of specialized facial products.
- **Manage your eyebrows.** Go to a spa or salon and get your eyebrows tweezed (or waxed), and, optionally, dyed a slightly darker or lighter shade.
- **Whiten your teeth.** Buy an over-the-counter tooth whitening system, such as Crest Whitestrips, and begin using it tonight. If you haven't seen a dentist in over a year, make an appointment.
- **Freshen your breath.** Start flossing daily. Consider getting a tongue scraper if halitosis is a problem. Buy gum or mints, and carry them with you at all times.
- **Get free dermatology advice.** Go to a department store cosmetics counter and ask the beautician what facial products she recommends for your skin type. Feel free to ask for samples or buy cheaper equivalents at a drugstore. If you consider your complexion to be a major liability, make an appointment to see a dermatologist.
- **Accessorize.** Buy a necklace, rings, a bracelet, a wrist cuff, or any other tasteful accoutrement. Try not to get anything that looks too cheap and mass produced—even if it is. When in doubt, err on the side of wearing something simple for now.
- **Join a gym.** Make an appointment with a trainer to get a fitness evaluation and exercise regimen that includes both cardiovascular training to reduce fat and resistance training to increase muscle mass. Make working out a borderline obsession.
- **Eat healthier.** Control your caloric intake and review your diet to limit saturated fats, refined sugars, excess salt, and food high in preservatives and carbohydrates. Eat fresh fruits, vegetables, and lean

protein. If you're more than forty percent over the weight you should be, consult a doctor about weight loss options.

- **Make sure your clothes fit.** Go through your closet and try on everything. If jackets drop off your shoulders, jeans droop off your butt, short sleeves stop at your elbows, or shirt necks hang down to your chest, either get the item tailored or donate it to a thrift store. Same goes for anything else that doesn't flatter you. Commit to replacing these items with well-fitting clothes that best suit your build.

If you have any grooming or appearance issue not listed above—be it underarm sweat, foot odor, an unsightly blemish, or your ex-girlfriend's name tattooed on your neck—this is the day to start taking care of it. Research solutions online; talk to fellow Challengers in the Stylelife forum; and, if necessary, pick up recommended products or make that doctor's appointment.

Don't let yourself off the hook when it comes to looks. You no longer have an excuse.

DAY

MISSION 1: Conquering AA

Today we're going to discuss the single most debilitating problem facing would-be Casanovas: approach anxiety.

Approach anxiety is a crippling disease that occurs when a man is confronted by the prospect of approaching an attractive woman. Symptoms include sweaty palms, increased heart rate, shortness of breath, and a lump in the throat. Psychologically speaking, it's less a fear of approaching than a fear of rejection.

If you hesitated before walking up to anyone during any of your field assignments so far, then you have approach anxiety. If you haven't been nervous yet, you probably will as the missions grow more advanced, or when you see that one special girl. It happens to the best of us.

So turn to your Day 6 Briefing while there's still time and read the cure proposed by Don Diego Garcia, a senior coach in the Stylelife Academy.

MISSION 2: If You Can't Say Something Nice . . .

Make sure you shower, shave, and feel good before you leave the house today. If you developed a confidence boosting ritual on Day 4, do it. If you purchased any new items yesterday, put them on. You're going out again.

Your mission: Give four women spontaneous compliments. Two of these women can be people you know—friends, coworkers, even your mother. But two should be strangers.

Avoid general compliments such as "You're beautiful." And avoid saying anything that could be construed as showing sexual interest, like "You're hot." Instead, focus on complimenting something specific, such as her nails, shoes, handbag, or posture. After spending time rigorously examining yourself yesterday, you should find it easier to spot and appreciate these details.

The most common response will be a sincere, polite, or dismissive thank-you. Leave after the compliment, unless she continues the conversation.

The key is to be perceived not as trying to flatter or hit on her but as showing sincere appreciation of something you've noticed spontaneously.

Though giving compliments isn't recommended for all approaches, generating attraction isn't the goal today. This exercise is designed to help eliminate approach anxiety, improve your skills of observation, and get you out of your head and aware of someone else's reality.

MISSION 3: The Eight-hour Rule

Get a good night's sleep, because tomorrow is one of the most crucial days in the Stylelife Challenge.

DAY 6 BRIEFING

ABOLISH APPROACH ANXIETY

By Don Diego Garcia

There are millions of words of wisdom offered by experts on creating and developing a successful intimate relationship, but seven words stand above them all: *You can't win if you don't play.*

That is the bottom line of bottom lines, courtesy of the California State Lottery. If you stay in your solipsistic cave, you will never form a new relationship. You *must* get out of the house and interact with new people.

Approach anxiety is a name for the internal demon that keeps men from talking to attractive strangers when there are no external barriers. Before working on ways to convert approach anxiety into approach excitement, let's discuss two key concepts: the limiting mind and the freedom mind.

The Limiting Mind

When we are born, nature installs two major instinctual fears to keep us safe: a fear of heights and a fear of loud noises.

Fear in moderation is a good thing. It protects us from harm. For example, a fear of heights protects us from falling off cliffs. A fear of loud noises enables us to react quickly to warnings of danger. However, most fears and limits we have are the result not of nature but of nurture. We place limits on ourselves as the result of negative experiences from our childhood and the influence of authority figures.

The Freedom Mind

The biological freedom mind gives us signals of hunger to eat, thirst to drink, and desire to procreate. In modern times, we also have cultural drives for power through career, enjoyment through play, and purpose through spiritual practice.

When our limiting mind and freedom mind are in homeostatic balance, all is good. We live in harmony with the world, effectively solving problems as they arise. But when our freedom mind and limiting mind fall out of balance, all kinds of afflictions arise.

Identify Your Limiting Mind

Most of your limiting mind's beliefs were spoon-fed to you by your parents, guardians, teachers, clergy, peers, or whomever you admired while growing up. While there is some value in tracing the sources of your own personal limiting mind, it's more important to understand its structure. The limiting mind tends to feed on itself in a downward spiral. Placing blame on others or on yourself for the material in your limiting mind only serves to strengthen it. It's best to forgive, forget, and move on.

The first step on most roads to recovery is acceptance—admitting that there's a problem. The second step in overcoming the source of our anxiety is to bring it out of unconscious darkness and into the light of our conscious awareness. Only then can we begin to dismantle it, see how it works, and create procedures to nullify it.

The limiting mind may present hindering voices, images, or physical feelings when it's time to approach strangers and make their acquaintance. Let's identify the types of internal media it can use to intimidate you into aborting a social mission.

Voices of the limiting mind include:

- **Self-doubt:** "You won't know what to say" or "Remember last time you messed up?"
- **Other-oriented doubt:** "She probably has a boyfriend," "She wouldn't be interested in me," or "She's busy and I'd be interrupting her."
- **Environmental doubt:** "Everyone around will make fun of me" or "It's too loud for her to hear me."
- **Existential rationalization:** "Why bother? It won't work out anyway," "I don't feel like it right now," or "I'm having too much fun with my friends."
- **False judgments:** "She isn't attractive enough" or "She seems way too shallow for me."

Images of the limiting mind include getting ignored; being mocked or bullied; being sad and alone; being observed and judged; getting beaten up; being rejected; and seeing more qualified or successful men in the room.

The limiting mind also expresses itself through physical sensations. When a potential threat registers on your radar, the acute stress response (also known as the fight-or-flight response) releases adrenaline into your system. This hormone increases your breathing and heart rate; constricts blood vessels; tenses muscles; dilates pupils; elevates your blood sugar level; and weakens your immune system.

Awaken Your Freedom Mind

To abolish approach anxiety, convince yourself logically that the dialogue of your limiting mind is incorrect and in fact self-sabotaging. In your Day 1 reading assignment, several limiting beliefs were disproven. These are the kinds of rational responses your freedom mind can use when the limiting mind rears its ugly head.

For example, if your limiting mind tells you, "She won't hear you," your freedom mind should answer back, "If she doesn't hear me the first time, I'll smile and politely repeat myself more loudly, slowly, and clearly."

If your limiting mind tells you that you're going to get nervous, your freedom mind can say, "I may have a natural stress reaction to this situation because, after all, it is somewhat stressful. But that doesn't mean I won't be able to push through it. In the past, nervousness has given me the energy I needed to perform at my best and feel good about myself. So let's do this!"

Take a moment to write down your own limiting mind's reservations about approaching. Then write down corresponding freedom mind responses that empower you. Use the word *you* for the scripts of your limiting mind, and the words *I* and *me* in your freedom mind responses. This will help you disassociate from your limiting mind and associate more closely with your freedom mind.

It's up to you to feed positive scripts into your freedom mind on a regular basis, to give it the power to overcome, persevere, and succeed. To do this, pick three freedom mind scripts or affirmations that you feel would best replace your specific fears, whether they're the ones you just wrote down or ones included in this book. Write them on a single sheet of paper. Then read them out loud with conviction during your morning or evening freedom mind ritual, and run them through your mind over the course of the day. Once you start to feel the beneficial changes, switch to another set of affirmations according to your new needs.

Shift Your Submodalities

Submodalities are the media through which your senses receive, remember, and process information. For example, auditory submodalities include volume, pitch, tempo, and timbre.

To help eliminate negative internal dialogue, try adjusting the submodalities of your limiting mind's voice. Make it quieter and further away; stammering and squeaky; or use the voice of a person you don't like.

At the same time, give your freedom mind a strong, low-pitched, calm, nearby voice. Consider making it the voice of someone you respect: a mentor, an actor, or your future best self.

If these exercises seem at first glance like New Age tripe, that's your limiting mind at work again. This process is exactly what trainers instruct top athletes to do to master their game. It's also one way that therapists eliminate phobias.

Visually, put your mental pictures and movies through the same filters. First, overpower the images of failure in your limiting mind with the successful images of your freedom mind. Change a picture of getting ignored to one of being adored; change a picture of being rejected into a bright, vivid visualization of a beautiful woman pressing her phone number into your palm.

Now change the submodalities. Make the images in your limiting mind small, distant, black-and-white, slow-moving, blurry, and dark. Disassociate with these negative images by seeing them not through your own eyes but as if you're watching yourself as a character on a movie screen.

Whenever your limiting mind images pop up, instantly replace them with large, bright, sharp, colorful pictures of successful situations. Associate with these images by seeing them through your own eyes.

These mental exercises are best done just after waking up or before going to sleep, because that's when your subconscious is most open to changework. By repeating this exercise as often as possible, you'll get to the point where you automatically reject the negative images your limiting mind tries to throw at you before each approach.

Let Go of Your Outcome

One of the biggest problems men have with approaching women is magnifying the meaning of the interaction and focusing too intently on achieving one specific outcome—whether it be exchanging phone numbers, making out, having sex, or beginning a romantic relationship.

Emotionally detaching from the outcome—while rationally working toward your goal—will significantly alleviate your anxiety. This is why the Stylelife Challenge offers small, easy-to-accomplish goals rather than large, unlikely ones.

People can be random, unpredictable, chaotic creatures. And sometimes you may truly be surprised. That's why approaching is so much fun. So why constrain the possibilities of a new encounter by being dependent on a particular outcome?

Remove Failure from Your Vocabulary

The word *failure* has different meanings for different people. To most people, failure means approaching and being rejected. My definition of failure is quitting, giving up, or never approaching at all.

Rejection is another word that's been misused and misrepresented. The dictionary definition of *reject* is "to refuse to accept." So if you offer someone a stick of gum, and she says "No thanks," you've been rejected. Do you feel an emotional sting? Probably not.

If you invite someone to a social event, and she says "No thanks," it shouldn't be any different. But for most people it is different, and here's why: When the gum is rejected, we think the person doesn't want the gum. But when we extend an invitation and get rejected, we think she doesn't want us.

But how could she possibly have decided she doesn't want us? She's known us only for a short while. She's practically a complete stranger. She doesn't know how great we are, the way our friends and family do. Why do we value her opinion over theirs? Why do we attach so much emotional baggage to a virtual stranger's ill-formed opinion? You guessed it: the limiting mind.

Practice the Crash and Burn Strategy

If, after reading this, you still have a crippling fear of social rejection, then go out and try to get rejected. Every accomplished social artist I know has a ton of rejections under his belt. That's simply the price you have to pay for excellence.

To quote Michael Jordan, "I've missed more than nine thousand shots in my career. I've lost almost three hundred games. Twenty-six times, I've been trusted to take the game-winning shot and missed. I've failed over and over and over again in my life. And that is why I succeed."

After a few rejections, you'll see that it's not so bad, that rejection really has nothing to do with who you are. It's more like somebody flicking you in the shoulder with a finger. You know it happened, but it doesn't hurt you or really even bother you. It's actually just immature and embarrassing on their part.

I took a student out once and tried to get us rejected to help him past his fears. But a funny thing happened: My plan backfired, and I wasn't rejected at all. The conversation went something like this:

ME: Hey! How are you doing? Could you blow us out? We need to get blown out.

THEM: Huh? What's that?

ME: Oh, that's when a couple of guys roll up and you're in some mood, so you're totally rude and don't wanna talk, and you tell the guys to—

THEM [*INTERRUPTING*]: Oh, we're not rude. Not at all!

We ended up having a pleasant conversation for forty-five minutes, after which we exchanged contact information. The exercise was supposed to demonstrate that blow-outs are pain free, but it ended up teaching a different lesson: that you can open by saying almost anything when you're confident, congruent, and upbeat.

Feel free to prove it to yourself. Next time you see someone you want to talk to, open your mouth and say the first thing that comes to mind. As long as your comment or question isn't rude or hostile, you may be surprised by how difficult it is to get solidly rejected.

After trying this a few times, you'll also notice that everyone's responses vary. Then you can adjust your attitude to expect nothing and prepare for everything. Or, as the poet Samuel Hazo puts it:

Expect everything, and anything seems nothing.
Expect nothing, and anything seems everything.

STOP!

DID YOU COMPLIMENT FOUR WOMEN?
DID YOU SHOP FOR NEW CLOTHES?
DID YOU CREATE YOUR MISSION STATEMENT,
DO THE POSTURE EXERCISE, GET MOVIE RECOMMENDATIONS
FROM THREE STRANGERS?

IF YOU ANSWERED YES TO ALL OF THESE QUESTIONS, THEN
PROCEED TO THE NEXT PAGE.

IF YOU HAVEN'T ACTUALLY BEEN DOING THE MISSIONS
BUT JUST READING TO GET THE INFORMATION,
THEN DO NOT PROGRESS PAST THIS PAGE UNTIL YOU CAN
ANSWER YES TO THE QUESTIONS ABOVE.

READING THIS WORKBOOK STRAIGHT THROUGH
IS LIKE GOING TO THE GYM TO WATCH TELEVISION.
YOU'RE NOT GOING TO IMPROVE
IF YOU DON'T DO THE EXERCISES.

DAY

MISSION 1: Learn to Open

Your first lesson today: There is no such thing as a pickup line.

If there were a single sentence that magically made women fall in love or lust, every man would be using it. Most of what people call pickup lines are actually comedic one-liners that were never legitimately used to meet women in the first place.

What does exist is a specific sequential process that can be used to develop a romantic or sexual relationship with a woman.

And this process begins with the opener, perhaps the most important part of the interaction.

Your task is to turn to your Day 7 Briefing and read the field guide to openers before beginning the next mission.

MISSION 2: Prepare Your Opener

Your mission is to develop an original opener based on today's briefing.

The simplest way to generate an opener is to think about anything you're curious about, want to learn, or are confused about. Choose a topic that is likely to capture the interest of most people. It can be a meaningful, debate-inspiring subject based on a relationship or spiritual crisis, or it can be a specific, trivial subject based on a popular culture, travel, health, or social customs query.

Then, instead of asking a friend about the subject or looking up the information on the internet, use it as a reason to talk to other people. For example, if you can't remember who sings a certain popular song, make it your mission when you leave the house today to ask strangers until you get a correct answer.

If your friend's girlfriend tried to kiss you, and you don't know whether to tell him or not, by all means, get some advice from the woman in the street.

Even unlikely questions can be effective openers as long as they're genuine. For example, I was having a debate with a friend one day over the names of the oceans. So, rather than seek the immediate gratification of Google, we made it our opener for the night: "Hey, how good were you at high school geography? Okay, how many continents are there? Right, seven. And how many oceans? Okay, five. So here's the question: What are the five oceans? My friend and I have been stuck on this all day. We can only come up with four."

As ridiculous as it sounds, it started a conversation every time.

Although today's briefing mentions different types of openers, for this task, focus on indirect openers that don't convey sexual or romantic interest. Make sure your attitude about whatever you ask is positive and that you avoid discussing anything that might reflect badly on you, such as creepy topics like serial killers or insecure questions about yourself.

MISSION 3: Test Your Opener

Get groomed, get dressed, and get excited. Your mission today is to approach three different women—or groups that include women—and deliver either an opener you've invented or one you read in today's material. You may approach in the street, at a café or bar, in the mall, in an office waiting room, or wherever you choose.

It isn't necessary to continue the conversation afterward, but feel free to do so if it's going well. When the discussion comes to a natural close, exit with a simple line: "Thanks. Nice meeting you," for example.

It is not necessary to have three successful interactions; just three approaches. Tomorrow we'll add a few extra pieces that will greatly increase the success and effectiveness of your openers.

MISSION 4: Evaluate Your Approaches

In the space below, make a list of the approaches you did today.

If any went well, write down the reasons you believe they worked. If any went poorly, make a note of why you believe they weren't successful.

Approach #1:

____

____

____

____

Approach #2:

____

____

____

____

Approach #3:

____

____

____

____

Now review your list. Do any of your reasons blame someone else for a negative outcome ("She was walking too fast," "She was stuck up," "She wasn't my type," "The guy she was with was an asshole")? If so, cross them out and replace them with an error you may have made. Then write down a suggestion for what you could have done differently to make the approach more successful.

DAY 7 BRIEFING
A FIELD GUIDE TO OPENERS

"What's your name?" "What do you do for work?" "Seen any good movies lately?"

Boring!

Listen to any man in conversation with a woman he's met, and chances are she'll be subjected to a nonstop barrage of questions that include one or all of the above. And because she's answering them, the guy will think he's getting somewhere.

Here's a question for you: How many times do you think she's answered those same questions before?

Answer: countless times.

Usually, the scenario ends like this: Slowly she starts looking around the bar, losing interest. The guy makes a desperate move and asks for her phone number. She politely says she has a boyfriend, even though she doesn't. Game over.

Why does this happen?

The comedian Chris Rock knows why. He has a routine in which he explains that anything a man says to a woman translates as "How about some dick?"

If you barrage a woman with generic questions, what she hears is "How about some dick?" Offer to buy her a drink, she hears "How about some dick?" Introduce yourself to her, comment on her necklace, ask for the time: "How about some dick?"

Your goal as a Challenger is to start a conversation with a woman without saying "How about some dick?"

This is accomplished through what are known as indirect openers. An indirect opener is a way to start a conversation with a stranger or a group of people you don't know without hitting on anyone or showing any romantic interest. If you do this well enough, soon she'll be asking *you* those generic questions.

The following guide includes the basics of using and developing these openers. Tomorrow, you'll learn two additional techniques to make them nearly failsafe.

Types of Openers

A successful opener serves four basic objectives:

- It's nonthreatening and makes no one uncomfortable.
- It stirs up curiosity and captures the person's or group's imagination.
- It's a springboard for follow-up conversation.
- It serves as a vehicle for you to display your personality.

There are many different types and classes of openers. These include:

- Direct openers, in which the man shows his romantic or sexual interest right away;

- Situational openers, in which the man comments on something in the environment;
- Indirect openers, in which the man initiates a spontaneous, entertaining conversation that is not about the woman or the environment.

All of these openers can work, but the first two often fall into the "How about some dick?" category. It's okay to use them, but only if the woman is initially interested in you or predisposed to be attracted to you. And even then they may not always work.

I prefer indirect openers because, when performed correctly, they work 95 percent of the time. And those are pretty good odds in this game, or any game.

Most indirect openers are premeditated and scripted. It may seem contrived and unnatural to prepare something to say, but when you have a conversation starter ready to go at any time, you don't have to hesitate and try to think of something clever to say every time you see a woman you find attractive.

Eventually you'll be able to start a successful interaction by spontaneously saying just about anything. For now, though, think of indirect scripted openers as training wheels—ones that work so well many guys never want to remove them.

Before the Opener

The game begins before you open your mouth.

Because the initial approach is such a critical moment, everything from your body language to your energy level takes on extra significance. Here are a few points to keep in mind when approaching a woman or a group of strangers:

- Always have something better to do than meeting women. As soon as you start staring at, evaluating, or ogling a woman in front of you, even if she can't see you, you've just lost every woman behind you. The reason is not just that you may seem creepy and desperate, but also that you don't seem interesting, fun, or worth meeting.
- Everyone wants to be with the most popular person in the room.

Since most groups in public settings don't know each other, all you need to do is create the illusion of being popular in that moment. From the second you walk in, be engrossed in an animated conversation with your friends. Smile, laugh, have fun, and enjoy one another's company.

- Then, when you notice someone you want to approach, wheel around and start a conversation. Don't hesitate or waste time assessing the situation. The art of the approach is the art of spontaneity. If you wait too long, either she'll notice you scoping her out and get creeped out—or, more likely, you'll think about it for too long, get nervous, and talk yourself out of approaching.
- Don't face the person or group head-on when you first approach. It's too direct and confrontational. Instead, turn your head and ask over your shoulder. Your goal is to give the impression that you're on your way somewhere else and just pausing briefly to ask some random people a quick question en route. Once the group begins to enjoy the conversation, you may turn and face them.
- Don't hover over or lean into the person or group. If you're competing with loud music or they're seated, just stand up straighter and talk louder. If all goes well, you'll soon be sitting down with them or moving somewhere quieter together.
- Smile when you approach. Even if a grin doesn't come naturally, fake it. It predisposes the woman or group you're about to engage to respond positively. On a subconscious level, it signals that you're a friend and not an enemy.
- Your energy level should be equal to or slightly higher than the woman or group you're approaching. Most people are out to have fun. So if you can add to their fun, you'll be welcomed into the group. If you're bringing them down or making them strain to understand you, it doesn't matter what you say—they'll want to get rid of you as soon as possible. Ways to increase your energy level include talking louder, using hand gestures, making an effort to connect with the people you're talking to, and smiling with your mouth and eyes. But don't be too hyper, because that's just annoying.
- Make sure that everyone can hear you, is paying attention, and is involved in the conversation. If you lose just one person, you risk los-

ing the whole group. So if you feel like someone's interest is waning, pull her into the conversation by addressing her directly or commenting on something she's wearing or doing.

- Don't be afraid to approach groups that include men. The more men there are in the group, the less likely it is that the women in it have been approached. You'll be surprised at how often the guys they're with aren't actually their boyfriends or husbands.
- Make sure you pay attention to the men in a group. If they feel you're not respecting or acknowledging them, they'll try to end the interaction. If you think any of the men mistakenly believe you're hitting on them, mention an ex-girlfriend or a crush on an actress.
- If you're interested in an attractive woman or group of women who've been hit on a lot, don't approach them directly. Instead, open a group next to them. Then, during a high point of the interaction, casually involve the woman you originally wanted to meet in the discussion.

What to Say

There are three traits a successful indirect opener should possess: It should appear spontaneous, be motivated by curiosity, and be interesting to most people.

There are also many subtleties. Never begin by asking a question that requires a yes or no response. If you say, "Can I ask you a quick question?" the group can always answer, "No." Then you're stuck.

Instead, begin with a statement, such as an observation, "You guys look like experts," or a request for assistance: "Help me settle a quick debate" or "Let me get your take on this." Then pause briefly to make sure you have everyone's attention, and continue.

Even when you ask your actual question, it's not necessary to get an answer. Pause for a moment, and if no one fills in the silence with an opinion, continue with your story.

Don't begin the opener by saying "I'm sorry," "Excuse me," or "Pardon me, but." Sure, your family raised you to be polite, but starting a conversation this way makes you sound insecure at best and like a panhandler at worst. Where men are initially attracted to beauty, most women are initially attracted to status. And a man of high status never apologizes for his presence.

The most widely used kind of indirect opener I've come up with is the opinion opener, in which you ask a group for advice on a personal story. A well camouflaged opinion opener can still evoke ten minutes of excited responses—which are also ten minutes you can use to showcase your humor and personality.

An easy opener for beginners is the "shady friend opener," which was based on a girl I dated. One bonus with this routine is that it can help you ascertain if the girl you're interested in is too jealous to seriously date.

Here's a word-for-word script. It was originally created in bars and clubs, so if you're out by yourself during the day, instead of pointing to a friend in the room, pretend you just got off the phone with him.

YOU: Hey guys, let me get your take on something. I'm trying to give my friend over there advice, but we're just a bunch of men—so we're not really qualified to comment on these matters.

THEM: What's that?

YOU: Okay, this is a two-part question. If you've been dating a guy for three months and he doesn't want you to hang out with one of your male friends, what's the appropriate response? Assuming that the person is just your friend, and nothing would ever happen.

THEM: I'd probably break up with the guy I'm dating.

YOU: Okay, here's the second part of the question. What if this friend was someone you used to sleep with? Does that change things?

THEM: Well, I'm friends with some of my exes, but others I can't be friends with. So it depends.

YOU: Okay, makes sense. The reason I'm asking is because my friend over there has been dating a girl for three months, and she wants him to stop talking to a female friend of his. He hasn't dated this other girl for years, and they're really just friends. The problem is, if he stops talking to her, he'll resent his girlfriend. But if he keeps talking to her, his girlfriend will resent him.

THEM: Something like that happened to me once, and . . .

If you're talking to a group, make sure you ask all the members—even the men—for their opinions. No one should be excluded, because if they are, they'll feel slighted or get bored—and could influence the group to shut you out.

Most important, as you deliver this or any other opener, remember that it's not the exact words that matter—it's your attitude. The opener is used only to break the ice and get the group's attention. It contains no magic formula that will make a woman swoon at your feet. It's just a way to keep your mouth moving while you display your charming personality.

After the Opener

A good opener will naturally lead to other questions and topics of conversation.

Often, you'll be asked for your take on the dilemma you've asked about. Make sure you have one. If you're normally a sarcastic or negative person, this worldview may create a bond with some women, but it rarely creates attraction. I know because I used to be that way, until I discovered that one of the keys to drawing people to you—and making them want to stay there—is radiating positivity.

This is why it's best to draw openers from your own life. If the opener is about someone in college, you should know what college it is. If it's about someone in another country, you should know what country it is. Determine in advance the ages, professions, relationships, and other details of the people in the openers you use. If you deliver the opener correctly, she will most likely be curious and ask follow-up questions. So be prepared.

But don't overprepare. You'll come up with plenty of clever responses to common questions, related topics to discuss, and interesting details in the moment. For example, if you're using the shady friend opener, and it elicits a flurry of conflicting opinions, you may find yourself saying, with a bemused smile, "You guys are great. You're just like *The View*."

However, beware of a common beginner mistake: milking the opener. As soon as the energy starts to flag, or you catch yourself thinking too hard of something to say to continue the conversation, the opener is over. Cut the thread and move on.

You'll learn exactly what to say next in future Challenge assignments, but for now just remember: As soon as you start struggling to keep a dying conversation topic going, you may as well be asking "How about some dick?"

The Rule of Trying

Now that you're learning scripted material, it's important to remember the rule of trying: *Don't*. If you try hard, you die hard.

As soon as you're caught trying to impress her, trying to get validation, trying for attention, or trying too hard in any way, the game is over. One of the paradoxes of the game is that it takes a lot of effort to appear effortless.

While it's possible that in the future certain routines and lines in this book may become well known, the principles upon which they work have always been and will always be true. So feel free at any point to go to www.stylelife.com/challenge to learn new and proven openers created by Challengers and coaches.

As you become more advanced, you'll find yourself relying less on prescripted openers. You'll eventually be able to go out with friends and challenge one another to come up with the most ridiculous opening lines possible. And as long as your attitude is upbeat, non-needy, empathic, and positive, you'll discover that you can do no wrong.

Troubleshooting

Tomorrow you'll learn the two keys to avoiding most things that can go wrong during an opener.

For now, just remember that whatever happens during the opener is feedback. A rejection is not a comment on you but on your technique.

If a woman tells you that she has a boyfriend (and you haven't asked), it means she thought you were hitting on her. If she says she has to go to the bathroom, it means you made her uncomfortable. Adjust your future approaches based on these responses and develop answers that will transform common objections into attraction-building material. For example, if she accuses you of using a pickup line, you can respond, "You thought I was hitting on you? That's cute, but I don't think you could handle me."

Whatever you do, always remember the golden rule: You must open.

If you don't approach, you'll never know whether that stranger could have become a girlfriend, a casual fling, a good friend, or even a career opportunity. Almost every student I've talked with has regrets about not approaching a girl. But few have ever regretted making an approach, no matter what happened.

The pain of letting yourself down is much greater than anything someone else can say.

DAY

MISSION 1: Fine-tune Your Openers

Congratulations on delivering your first openers. Some of you may have found that conversations began with ease. Others, not so much. If you felt like you were bugging people, if someone asked whether you were taking a survey, or if you got funny looks, that doesn't mean you did anything wrong. It just means you're ready for your next mission.

Today you're going to learn two key subtleties of opening. Once you add these pieces to your approach, you'll notice a big difference in the effectiveness of the opener and the responses you get.

So turn to your Day 8 Briefing and read about the two keys before continuing to the next mission.

MISSION 2: Approach with Your New Tools

Approach three women—or groups that include women—with the opener you used yesterday.

This time, add both a root and a time constraint to each approach.

MISSION 3: Evaluate

When you return home, ask yourself if there was anything different about the responses you received from women you approached today, compared with those you approached yesterday. List three differences in the space below:

____

____

If you used an opener you made up, but it didn't seem to spark a natural conversation, then in future missions try using one of the scripts provided in this book (such as the shady friend or five oceans openers), or examine and modify your opener.

If you're not sure whether your opener is effective, post it on the Stylelife website message boards. There your fellow Challengers will evaluate and, if necessary, strengthen the material.

DAY 8 BRIEFING

THE TWO KEYS

As soon as you approach a group of strangers, they generally think two things: "What does this person want from me?" and "How long is he going to stay here?"

One of the strategies of the game is anticipating and defusing these objections—and any objections—before they happen. If you do this successfully in the first minute or two of your approach, you'll be much less likely to receive negative or flat responses.

Rooting

If a woman doesn't know why you're talking to her, she'll generally be suspicious until she either finds out from you or guesses her own reason. This is why people using opinion openers for the first time are often asked if they're taking a survey.

To anticipate the question "What does this person want from me?" you need to "root" your opener by giving your question a legitimate context.

For example, the opener may be something that's just now come up in your life, and there's a slightly urgent need to get an answer immediately.

The best way to convey this is to explain at some point during the opener why you're asking. You can use the following words to introduce your root: "The reason I'm asking is because . . ."

In the shady friend opener, the reason you're asking is that your buddy just moved in with his girlfriend, and she doesn't want him to talk to one of his female friends. And you were just now trying to give him advice, but he won't listen and you need some backup.

The root doesn't always need to be elaborate. It can be as simple as: "My friend and I were just talking, and we need a woman's perspective." If you're not with a friend, then it can be a discussion you were just having on your cell phone. Anything reasonable qualifies as a root, as long as it lets the woman or group know why you walked up and started talking to them about that particular subject at that very moment.

Time Constraints

For most inexperienced men, the game consists of approaching a woman and trying to stay in constant conversation until she either dismisses him or sleeps with him. Because of this, women have developed a vast array of tactics to get rid of guys who lurk too long.

This is why, from now on, you're going to let her know right away that you're not one of those guys. Unless she's already attracted to you, from the minute you approach she will most likely be wondering how to get rid of you. Her strategies for doing so may include telling you she's in the middle of an important conversation with her friends, claiming she has to go to the bathroom, or pretending that she has a boyfriend or is a lesbian.

So to anticipate the question "How long is he going to stay here?" you'll need to use a time constraint.

A time constraint is anything that explicitly lets the woman or group know that you don't plan on hanging around long. It should be inserted in the first minute of conversation, before the group has the chance to wonder when your story is going to end. So preface the opener you've been using with a time constraint like, "I have to get back to my friend in a minute, but, really quickly . . ." Or, in the middle of your opener, explain, "By the way, it's guys night out and I shouldn't even be talking to you all."

A time constraint doesn't have to be verbal. It can be physical as well. This is conveyed by leaning away, rocking on your back foot, taking a few steps away as you're talking, or anything else that makes it look like you're in a hurry or on your way somewhere else.

The best time constraints contain both elements: They're expressed verbally and sold through body language.

When you use both a time constraint and a root, it allows the woman or group to stop worrying about what you want and how to get rid of you, and relax enough to listen to what you have to say.

But wait, you may be thinking. If you just told her you have to leave in a minute, how are you supposed to keep talking to her after the opener?

Good question.

The next key stage of the interaction is known as the "hook point." This is when, instead of being a stranger taking up her time, you've captivated her—and suddenly she doesn't want you to leave. So, reluctantly, you allow her to take up a little more of your precious time.

Becoming that guy is what the next week of the Stylelife Challenge is all about.

DAY

MISSION 1: Crunch Time

Next week, the pace is going to pick up. So to make sure you're caught up and ready to proceed, today is review day.

Your task is to look over the previous eight days of assignments. Then ask yourself:

- Is there any mission I skipped?
- Is there any mission I feel I didn't complete?
- Is there any mission I didn't perform to my satisfaction?
- Is there any mission I'd like to do again?
- Have I backslid in my vocal training, posture, grooming, or commitment to my goals?

Take this opportunity to explore or repeat any previous assignments and exercises you need to reinforce.

MISSION 2: Approach Mixed Groups

If you've approached only lone women or groups of women during the Challenge so far, then it's time to approach groups that contain men.

Your mission is to approach two groups of three or more people that include men as well as women.

Approaching groups with men may sound daunting if you haven't done it yet, but it's generally easier in practice. The more intimidating people are to approach, the less likely it is they've been approached.

Don't forget, all you have to do to ensure the success of the approach is make sure that the guys are always involved in the conversation, they feel respected, and they know you're not hitting on the women. At least not yet.

MISSION 3: Intervention

Statistically, the ninth day of a new self-improvement program is the point when most people drop out. That's not going to be you. So your final task today is to read your Day 9 Briefing and prepare to learn how to learn.

DAY 9 BRIEFING

THE FOURTEEN LAWS OF LEARNING

When I first set off on my journey to learn the game, a college junior named Chad emailed me. He had discovered the world of pickup artistry six months earlier and was already well versed in the basic concepts. However, he was still a virgin.

He was far better looking than I was, with a stocky build, wavy black hair, and a square jaw. Yet a year later, I was having fantastic adventures that I'd never thought were possible for a guy like me. And Chad, despite working just as hard, was still a virgin. So I sat down with him one night and tried to figure out why. The reason, we eventually realized, was that we had different strategies for learning.

Afterward, I began developing the fourteen laws of learning that follow. They apply not just to the game, but to school, work, and hobbies. They are what separate a chump who's banging his head against the wall in frustration from a champ who's smoothly ascending to the top of the game. Make sure you understand and can practice each principle before moving on to the next.

1. **Acquire and apply knowledge in small chunks.** Some people are perfect preparers. They want to gather every scrap of information on a subject before taking action. And though they seem to be working hard, this is actually a form of procrastination. The best way to learn the game is to take it one step at a time. Just learn what you need to get to the next level. If you can't approach women, just work

on openers. When you master openers, then learn how to continue the conversation. Don't worry about advanced sexual techniques. You'll soon get there if you continue to progress by adding one piece at a time as you need it.

2. **There is no such thing as rejection, only feedback.** A lot of people get discouraged and give up after a single setback or rejection. They tend to take rejection personally, seeing it as a comment on who they are rather than what it really is: feedback on what they're doing. Every time you approach a group of people and something goes wrong, you've been presented with an opportunity to learn why they responded negatively and what you could have done to prevent that. If you possess the ability to learn from your mistakes, then failure is literally impossible, because each rejection brings you closer to perfection.

3. **It's never her fault.** Who do you blame when something goes wrong during an approach? If you catch yourself saying that a situation was impossible, the guys were jerks, or the woman was just a "bitch," then you're wrong. It was your fault. It's always your fault. And that's a good thing, because it means you're in control. So never blame any person or situation. Instead, demonstrate a willingness to examine yourself and accept criticism *without taking it personally.* Only then can you accurately determine whether there was something you could have done to change the outcome, or if the outcome was truly unavoidable.

4. **Learn actively rather than passively.** Just as you can't learn to play football by watching videos and posting in football newsgroups, the only way to learn to attract women is from real-world experience. Anyone can sit in a seminar or buy a DVD and learn the principles, but the guys who win the game are the ones who can apply them.

5. **Don't rehearse negative outcomes.** One of the biggest problems men have when it comes to meeting women is that they re-

hearse negative scenarios in their minds. Often, these become excuses not to go out and try something new. Instead, get out of the house, make a few approaches, and if any of these scenarios happens to occur in real life, *then* find out what to do. This isn't skydiving: There's little to no risk of actual harm from being unprepared.

6. **Understand how your mind learns.** The psychological field of neurolinguistic programming (NLP) offers a useful four-step model of how the mind learns. It can serve as a yardstick to measure your progress.
    - *Unconscious incompetence:* You're doing something wrong, and you don't even know you're doing it wrong.
    - *Conscious incompetence:* You're doing something wrong, and you're aware that you're doing it wrong, but you haven't yet fixed the problem.
    - *Conscious competence:* You've learned the right way to do it, and you're doing it correctly with focused attention.
    - *Unconscious competence:* You no longer have to think about something or work on learning it—you automatically do it correctly. In the parlance of the game, this is when you finally become a so-called natural.

7. **Be willing to go through the pain period.** This game is not an easy one. You'll be forced to confront nearly every single thing that defines you—every emotion, every action, every belief. You'll sometimes be apprehensive about approaching a particular woman, trying a new technique, or changing a behavior. What separates an amateur from a champion is the willingness to push through that fear and do it anyway. Here's what Arnold Schwarzenegger, in his iron-pumping days, had to say about it: "If you can go through the pain period, you make it to be a champion. If you can't go through it, forget it. And that's what most people lack: having the guts—the guts to go in and just say . . . 'I don't care what happens.' "

8. **Don't look to friends or family for approval.** Not all of your friends and family will understand the journey you're about to take.

They may tell you that they don't like how you're changing. They may make fun of you for wanting to improve. That's okay. It happened to me. It also happened to Oprah: When she lost weight, she lost friends. This surprised her at first, until she learned that her largeness had given them an excuse to feel better about their own bodies. So, when you start attracting women and adventure, your friends may not welcome it—you've become a threat to their limiting beliefs and complacency about their own shortcomings. Let it be their problem, not yours.

9. **Be willing to test new ideas, even if they don't seem logical.** Before I learned the game, I considered myself an intelligent and successful person. Yet the logic that had gotten me so far in the world wasn't getting me anywhere with women. In order to make a change, I had to try some new behaviors, even if they didn't seem logical. I said things I thought would drive women away, but instead they attracted them. I wore outrageous clothes I thought would get me laughed out of the room, but instead they motivated women to approach me. And that's when I realized that I'd never really been using logic in the first place—because, as any good scientist knows, before dismissing a new hypothesis, it's necessary to test it first.

10. **Once something works, figure out how and why it works.** There are some men who do great just following these instructions and repeating the routines. But the ones who become superstars are the ones who, after a series of successes, figure out *why* the routines worked and what made them work. There's only one rule of pickup, and that rule is: There are no rules, only guidelines. Once you understand the principles behind each idea, you'll know when to follow the guidelines, when to dismiss them, and when to invent new ones.

11. **If you don't know what to do, don't leave.** If you run out of material when talking to a woman you've just met, you're not going to learn anything by running away. Stay in the conversation and, if you run out of things to say, push it five, ten, twenty minutes fur-

ther—even if you have to violate the guidelines and buy her a drink or ask interview questions. It's the best way to learn something new for next time.

12. **Hang around someone better than yourself.** This is the single best way to improve in any area. Your mentor doesn't have to be the top attraction expert in the world, just someone who has a little more skill than you do. If you don't know anyone who can fill this role, instead of going out to meet women one night, go out to befriend someone who's good with women.

13. **Make sure that your ratio of effort to results is increasing.** When learning a new way of doing something, most people get worse at the task before getting better. That's normal. But you'd be surprised by the number of people who keep putting more work into something after this transition period, even though their results stay the same or barely improve. So make sure you're increasing not just your knowledge but also your results. If you're not, then take a break, review these rules, examine what you're doing, and push yourself beyond your comfort zone.

14. **Finish what you begin.** Most people can accomplish just about anything within the realm of possibility. Despite this, they never realize their dreams. Either they quit before they reach their goals (and always with a seemingly good reason for doing so), or they don't change their strategy when something's not working. Roughly 19 out of 20 people who start reading this book won't stick with the program until the end. Don't be one of those people. Simply by not giving up, you'll already be in the top 5 percent of men out there.

DAY

MISSION 1: It's Opposite Day

The focus of today's lesson is disqualification—one of the most counterintuitive techniques in the Stylelife Challenge. Forget everything you know about attracting women, because the goal of disqualification is to meet women and tell them you *don't* want to date them.

This is going to be the most difficult day of the Challenge so far—but also the most rewarding. To find out what it's all about, read your Day 10 Briefing and fill out the worksheet describing your ideal woman.

MISSION 2: Play Hard to Get

Your mission today is to make three approaches using one of the openers you've learned or created.

During the first approach, add a disqualifier from today's reading material.

For the second approach, use a different disqualifier.

Afterward, take a short break and think of a third potential way to disqualify her. Write it below:

____

____

____

____

Now make your third approach and, during the opener, use the disqualifier you just invented.

DAY 10 BRIEFING
THE POWER OF NO

It's not the having, it's the getting.
—ELIZABETH TAYLOR

I recently went to a party in Colorado with six friends. Three of the guys spent the night with women; three didn't. As we discussed it the next morning, we discovered that the difference between the unsuccessful guys and the successful guys boiled down to one thing: lack of neediness.

The guys who went home alone were too available. The successful guys all played hard to get. They weren't afraid to walk away from the woman they were attracted to, talk to other people at the party, and create the impression that if she didn't act soon, she'd lose her chance. They understood a basic tenet of human nature: The harder we have to work for something, the more we value it.

Thus the lesson for today: In every interaction, be the person *giving* validation, not the one needing it.

One of the quickest and most playful ways to accomplish this is through disqualification. To disqualify a woman, demonstrate early in an interaction that you're not interested in her. Even though you may be chasing her, disqualification turns the tables and makes her want to chase you. For example, telling a woman with blonde hair that for some reason you've only dated brunettes disqualifies her as a potential girlfriend.

If the concept sounds odd, consider this: Beautiful women are constantly approached by men. They assume that nearly every guy wants to sleep with them. So when you take yourself out of the dating pool in a confident way, you immediately stand out—after all, most people want what they can't have.

Another advantage is that disqualifying a woman in a group can help you win over her friends, who are used to repelling the steady stream of men vying for her attention.

Finally, disqualification helps build trust because it demonstrates that you're not solely motivated by the desire to sleep with her. By waiting before showing interest, you give her an opportunity to win you over with her charm, personality, and intelligence.

Not every relationship requires disqualification. Sometimes the feelings are mutual, and two people are attracted to each other right away. Also, if you're dealing with a woman whose confidence in her appeal is very low, you may want to avoid teasing her, since she's constantly disqualifying herself in her mind anyway.

Once you get comfortable using disqualifiers, you'll realize that they're not such a foreign, complex, counterintuitive concept at all, but in fact the bedrock of flirting.

Most disqualifiers are meant to be playful. Others are used to demonstrate that you have high standards and won't date or sleep with just anyone. However, a disqualifier should never be hostile, critical, judgmental, or condescending. There's a fine line between flirting and hurting. And disqualification is never intended to be mean or insulting. So say these with a smile on your face and laughter in your voice, as if you were good-naturedly picking on a younger sibling.

Screening

Women test men. They do so for many reasons: because they want to select the best potential mate from among many suitors; because they've been hurt in the past and don't want to make the same mistakes again; because they want confirmation that you authentically possess the qualities that attract them. Throughout your interactions with most women, whether they're consciously aware of it or not, they're putting you on the spot to see how you'll react.

These tests range from flirtatious teasing (such as telling a man he's too young or too old for her) to serious interview questions (such as asking a man why he and his last girlfriend broke up). Men normally sit there answering the questions like they're on a game show, hoping that if they accumulate enough points, she'll choose them. What they don't realize is that they're losing points simply by submitting to the test.

Screening allows you to flip the script and see if the woman you're interested in meets *your* standards. Before doing this, it's important to know exactly what your standards are.

Take a moment to imagine your ideal woman. Then list below five specific criteria you would like her to possess. Consider such qualities as personality, looks, upbringing, values, interests, knowledge, and life experience.

1. ______________________________
2. ______________________________
3. ______________________________
4. ______________________________
5. ______________________________

Now list five deal breakers. Qualities that might prevent you from dating someone could include manipulativeness, narcissism, smoking, drinking, drug use, jealousy, pets you're allergic to, and emotional baggage.

1. ______________________________
2. ______________________________
3. ______________________________
4. ______________________________
5. ______________________________

Keep in mind that this is just an exercise. When dating, remain open to the unexpected. If you're looking for someone who fits this bill exactly, you might overlook an even better match when she appears but doesn't meet your preset criteria.

In the meantime, this list will provide you with endless criteria for disqualification. On the simplest level, you can ask what her favorite films are and then act as if her answer is a deal breaker. "You actually liked that? That's it. I'm going home. Nice meeting you."

If you want someone who's adventurous, ask her: "What's the wildest, craziest thing you've ever done?" When she answers, disqualify her by saying, with a smile, "That's great. You and my grandma would really get along."

There's an endless list of potential criteria to screen her on, from her dancing skills to her preferred ice cream flavor to her lack of an Olympic gold medal (because you only date women with Olympic gold medals, so she'd better hurry up and get one).

The point of screening is never to make a woman feel bad about herself but to set yourself apart from the hordes of men who will sleep with anyone indiscriminately.

Push-Pull

The opposite of disqualification is qualification, or acceptance. When used together, these two techniques are very powerful.

If she says or does something good, give her a positive, accepting statement ("I like your attitude"); if she says something that could be perceived as negative, tease her with a disqualifier ("Note to self: Do not date this girl").

Taking control of an interaction by alternating back and forth between these two poles—punishment and reward, validation and invalidation, approval and disapproval, qualification and disqualification, push and pull—is one of the key ways to amplify attraction.

Like everything else in the game, push-pull should be doled out humorously and not cruelly. One way to make the process fun is to put her on a point system: Give her points for good behavior and subtract points for bad behavior. If you want to push it further, tell her that she can claim rewards at certain point thresholds: At forty points she gets to touch your bicep, at eighty she gets the first three digits of your phone number.

Perhaps the most fun form of push-pull is inventing a relationship prematurely. Tell her with a laugh that you're going to make her your girlfriend—on Fridays only—or joke that you're going to marry her on the spot. Then, moments later, pretend to be upset by something she just said or did and change the status of the relationship. Tell her you're demoting her to your Tuesday girlfriend, or you're filing for divorce and she can keep the cat.

10 More Ways to Disqualify

Disqualification can take myriad forms. Here are a few more to help with today's field assignment.

Remember, if you say these with a smile and a sense of humor, you'll come off as a great flirt. If you say them seriously, or as though you mean it, you're just an asshole.

- *Save her from you.* Often, trying to drive someone away is the best way to get her to chase you. Tell her you're the kind of guy her mother warned her about. Or say, "A good girl like you should probably be talking to a nice boy like that one over there." Not only does this make you seem fun and dangerous, but it inspires her to live up to that reputation as well.

- *Give yourself a monetary value.* This can be done by pretending it's a privilege to talk to you or touch you. If she takes your hand, pull it away and joke, smiling, "Hey now, hands off the merchandise. That'll be forty dollars."
- *Put her in the friend zone.* This is something women often do with men, but men rarely do with women. It can be done flirtatiously (by telling her she's like the little sister you never had), or more seriously, by telling her she'd make a great friend.
- *Go over the top.* Exaggerate her greatness and pretend to be an awe-struck admirer. If you say this in a wry, superior way, you'll actually end up conveying the opposite.
- *Reverse roles.* Everything she doesn't want a guy to do, jokingly accuse her of doing to you. Tell her to give her obvious pickup lines a rest, to stop treating you like a mindless piece of meat, to quit trying to get you drunk and take advantage of you because you're not that kind of guy. The more unlikely the scenario, the more effective your accusations.
- *Employ her.* Jokingly offer to hire her as your assistant, your web designer, or some other job she'd never do. Then, of course, fire her moments later.
- *Be the snob.* All those immature things the popular girls in school may have said to you, you may now say to her. Examples include: "Uhh, whatever," "Not so much," and "Yeah, you would say that."
- *Be the authority figure.* The annoying things your parents and teachers told you are also fair game. Playfully tell her she's starting to get on your nerves, she's in big trouble, or she's just earned herself detention.
- *Make her compete.* Threaten to leave to talk to your friends, the waitress, or those "more interesting girls over there."
- *Challenge her:* Tell her you're not sure yet if she's cool enough, adventurous enough, or mature enough to hang out with you.

The list is endless. Any line a guy might use to hit on her, you should say the opposite. And anything she might say to a guy who's hitting on her, you can say to her instead.

It's that easy.

Performance Notes

For most of you, disqualifiers won't come easy—not because they're difficult, but because they go against everything you've been raised to say around women you like.

Tone is everything. Except for when you're actually screening someone to see if she meets your relationship criteria, most disqualifiers should be delivered playfully. If you appear serious or upset when you accuse her of hitting on you or not being cool enough for you, she'll think you're a psycho.

Most disqualifiers should also be delivered casually and offhand, as if you're not seeking or expecting a reaction. If it's obvious you're just using the disqualifer for effect, it loses its power and becomes just another form of neediness.

Though being rich, successful, and good-looking is normally a good thing when it comes to the game, it isn't with most disqualifiers. The point of the disqualifier is to raise your status to her level or above. But if she thinks your status is already far above hers, then most of these comments will make you sound obnoxiously arrogant rather than playfully cocky. So evaluate the situation before getting too hardcore with the material.

Finally, if you dish it out, be prepared to take it. She may respond to your disqualifier with a sharp comment of her own. If she does, don't panic. This a good thing. It's called flirting. Just be prepared with an even more clever retort to fire back. If you're stuck for an answer, just nod your head, smile, and say, "Respect," as if she's met your approval.

DAY

MISSION 1: Refine Your Identity

Today we're going to focus on the most important piece in the game: you.

In nearly every successful approach, at some point you'll be asked what you do. If you've mastered disqualifiers, your initial response will probably be to tease her for asking "interview questions" and then to claim to be a professional hopscotch player. If she persists, however, you're going to have to answer truthfully, or else she'll think you're hiding something.

The work question is an opportunity that most people waste. One student used to answer, "I'm an engineer." Engineering, of course, is a noble pursuit, but he felt like it made him sound boring to women.

When I asked him what he was working on, he said he was going to school to learn about new mobile phone technology. So we developed a better way for him to answer the question. Now, when women ask him what he does, he responds, "I'm designing the mobile phone of the future."

Same occupation, different identity.

In your Day 11 Briefing, there's an exercise that will help you refine your identity and articulate what you do in a crisp, compelling manner. Your mission is to fill it out and learn to succinctly express what makes you special without bragging.

MISSION 2: Approach and Continue

Approach groups of three or more people that include at least one woman. Use an opener that contains a time constraint and a root.

When you're finished with the opener, continue the conversation by adding the following movements and lines:

1. Pretend you're about to leave, but take no more than one step away.
2. Look back at the group and ask, out of curiosity, "Hey, how do you all know one another?"
3. Be ready to respond with a question or comment. It doesn't have to be anything clever or complex. If they say they're friends from work, ask, "So where do you all work?" If they say they're related, say, "That makes sense. I wonder which one of you is the black sheep."
4. You may now leave if you wish, with your all-purpose closer, "Nice meeting you."
5. You may also choose to continue talking to the group if the conversation is going well. If anyone asks what you do, answer with the identity statement you created today. Try to use the statement in at least one of your interactions.

The task is complete after you have followed steps 1 through 3 with three different groups of people.

MISSION 3: Master Your Inner Game

Too many of us have no idea what goes on inside our own heads. We don't understand our emotions, our passions, our frustrations, our needs, our thinking patterns, and why we sometimes act the way we do. And even when we do understand these things, we often find it difficult to change them.

One of the best books on this subject is *Mastering Your Hidden Self: A Guide to the Huna Way,* by an ex-marine named Serge Kahili King.

Though I recommend reading the entire book, for today's assignment I asked Stylelife senior coach Thomas Scott McKenzie to prepare a report summarizing its application to attraction. If your inner game needs a new set of rules, this document just may change your life.

DAY 11 BRIEFING
IDENTITY WORKSHEET

1. What are your primary jobs, hobbies, and/or courses of study? Answer based on how you actually spend your time, not on what you think will please women.

____

____

____

____

____

2. Which of the items you listed above best defines you?

____

____

3. What are the most interesting or adventurous aspects of the job, hobby, or course of study you selected? List each aspect, along with the ways it could affect people.

____

____

____

____

4. Now imagine you're a recruiter for the job, hobby, or course of study you selected. Using the template below, prepare an advertisement to attract people who aren't involved in the field and know little or nothing about it. Your goal is to make the job or hobby sound important and exciting.

Become a ___________________________________
SELECT NAME OF JOB OR HOBBY

and you can ________________________________
INSERT YOUR AD LINE

Examples: Become an engineer, and you can design the mobile phone of the future. Become a guitarist, and you can tour the world playing rock shows. Become a web designer, and you can help with the images of the world's biggest corporations.

5. Now examine the ad line you wrote. Remove adjectives, adverbs, and other unnecessary hype words (such as "exciting," "biggest," "best," "most powerful"). Examine the verbs you used, and make sure they're exciting and active ("create" is better than "have"; "launch" is better than "do"). Then, using these tips, rewrite your ad line as simply, factually, and powerfully as possible in ten words or less.

   ____

   ____

   Example: "help with the images of the world's biggest corporations" could become "reinvent the images of corporations" or even "reinvent the images of Fortune 500 companies."

6. Rewrite your answer to question #5 in the first person (begin with the word "I").

   ____

   ____

   Examples: I reinvent the images of Fortune 500 companies.
   I'm designing the mobile phone of the future.

7. This is your identity statement. Say it out loud until you're comfortable with it. If you feel it's uninteresting or inaccurate, rework it until it feels right—or repeat this exercise (starting with question 3) until you have an identity statement that is both truthful and interesting.

Troubleshooting

Most of the guidelines of the game are based on perceived relative status, and they change depending on how she feels your status compares to hers at any given time. So if you currently have a high-status position in society, rather than playing it up, play it down. Do exactly the opposite of what's suggested above. Keep it vague. For example, instead of telling her you're the head of a major film studio or an award-winning screenwriter, just say that you "work in movies" and let her wring the details out of you if she so desires.

MASTERING YOUR INNER GAME—A BOOK REPORT

By Thomas Scott McKenzie

A man is but the product of his thoughts.
What he thinks, he becomes.
—MAHATMA GANDHI

I am a star. I'm a star, I'm a star, I'm a star.
I am a big, bright, shining star.
—DIRK DIGGLER, BOOGIE NIGHTS

It's been proven time and time again: Confidence is attractive. Confidence earns the admiration of your coworkers, the respect of your friends, and the interest of women. In fact, it's safe to say that without confidence, all the seduction techniques known to man will not help you attract the women you desire.

But many men struggle with this most crucial of characteristics. Difficult childhoods, less-than-model looks, meager bank accounts, dead-end jobs, piece-of-shit cars, receding hairlines, underarm odor, and dating dry spells all reduce worthy men to nervous, timid mice. Even men with rock-hard abs and shiny red convertibles are sometimes unable to look women in the eye and speak with a strong voice, because a domineering mother or ex-wife damaged their self-esteem and confidence.

Mastering Your Hidden Self: A Guide to the Huna Way by Serge Kahili King offers an antidote to these confidence poisons. King teaches that we are not helpless victims vulnerable to our mind's tyranny. Instead, *we* control our minds. We control our emotions. We control our perceptions, our feelings, and our outlook. Harnessing ancient systems, King offers a concrete way to reprogram your mind so that you can stride through life with confidence, energy, and power.

INTRODUCTION TO HUNA

In addition to the widely accepted teachings of the world's great religions and philosophies, a more esoteric body of secret knowledge has been shared by initiates throughout history. Building on both the mundane and the arcane, Huna offers a system of self-improvement that cuts through the confusion of modern life.

Essentially, Huna states that you are in control of your life, your mind, and your reality. "The most fundamental idea in Huna philosophy is that we each create our own personal experience of reality, by our beliefs, interpretations, actions and reactions, thoughts and feelings," King writes.

A corollary to this is that our creative potential is unlimited. "You can create, in some form or another, anything you can conceive," King continues. This is why it's important to replace limiting beliefs based on past dating experiences with unlimited beliefs about the present and future.

Within the Huna belief system, there are seven main principles.

1. *The world is what you think it is:* The foundation of Huna, this principle asserts that you create your own personal experience of reality. "By changing your thinking, you can change your world," King writes.

2. *There are no limits:* There are no true boundaries between you and your body, you and others, or even you and God. The divisions that we generally recognize are arbitrary constraints placed by limited consciousness.

3. *Energy flows where attention goes:* When you dwell upon certain thoughts and feelings, you write the plotline for your life. Focus is the fuel for your positive or negative perceptions. So, for example, don't give some girl who ignored you the power to ruin your day by letting yourself dwell on the incident.

4. *Now is the moment of power:* At this moment, you are not hindered by any past experiences, and you are not obligated to any future duties (except paying taxes, of course). "You have the power in the present moment to change limiting beliefs and consciously plant the

seeds for a future of your choosing," King writes. "As you change your mind, you change your experience."

5. *To love is to be happy with:* People exist through love, King says, and acknowledging this allows you to exist in a state of happiness with yourself as you are now and as you will become in the future.

6. *All power comes from within:* If you want to change your reality, you can't wait for divine intervention. It's up to you to change your existence. This principle also contains King's crucial admonition that "no other person can have power over you or your destiny unless you decide to let him or her have it." For some, this means that it's time to stop blaming friends, family, work, or society for holding them back from social success and start accepting responsibility.

7. *Effectiveness is the measure of truth:* Sit in any courtroom, and you'll realize there are many versions of the truth. In an infinite universe, King writes, there is no absolute truth, only "an effective truth at an individual level of consciousness." Put simply, do whatever works for you.

THE DETRIMENTAL EFFECTS OF NEGATIVITY

To improve your inner game, it's vital that you recognize the detrimental effects of negative thoughts and energy. "Generally speaking, negative attitudes produce inner stress, which translates to physical tension and can affect organs and even cells," King writes.

The simplest way to change a negative attitude to a positive one is to be aware of bad thoughts when they appear, then consciously change them to a positive opposite. "You can do this whether or not the apparent facts of the situation seem to warrant it," King adds.

THE SUBCONSCIOUS MIND

When it comes to the subconscious, the common perception is that it lurks in the recesses of your mind, never to be known until you spend years on a therapist's couch, only to discover that you're a helpless victim of some random childhood event.

King disagrees. He explains that we can, in fact, control our subconscious. "The subconscious is not an unruly, rebellious child, nor does it ever work against your best interests . . . Whenever the *ku* [subconscious] seems to be opposing you, it is because it is following previous orders that you either gave it or allowed to remain."

A good example of how you can train your subconscious involves changing habits. Mental and physical habits are learned responses stored in your subconscious memory and released by associated stimuli. Huna teaches that the only way to eliminate a bad habit is to give your subconscious a more effective way to deal with the stimuli.

One strategy is to consider changing your speech habits. Maybe you litter your speech with brain farts and pausers. At some point in your life, perhaps these pausers allowed you extra time to choose your words. Eventually, they became a habit. Instead of accepting this bad habit or trying to quit cold turkey, Huna teaches that we must replace it. "The important point here is that there is no vacuum in the subconscious," King writes.

So instead, teach your subconscious to dump your pauser by learning to speak more slowly. Or train yourself to tap your finger against something every time you have the impulse to say "um."

Your subconscious wants to help you. It's just that sometimes the subconscious gets poor training. "Your subconscious never works against what it believes are your best interests," King writes. "Unfortunately, the assumptions on which those beliefs are based may be very faulty."

By interacting with your subconscious, King argues, you can understand your motivations and change the ones that aren't effective. He provides several strategies for interacting with your subconscious.

First of all, King suggests that you give it a name. Next, you can try one of two forms of memory search. The first is called a "treasure hunt." For this activity, simply talk to your subconscious as though you're chatting with a new pal. Name a memory of something pleasant and see what the subconscious brings back in terms of detail and vividness. Or you can ask your subconscious to return its own favorite memories. Memories you had forgotten will appear, and sensations will come flooding back.

The second form of memory search is called "trash collecting." For this activity, ask your subconscious to bring up all its worst memories. Do this enough, and you'll begin to see patterns. "The memories will follow certain themes that will provide you with clues to areas of limiting beliefs that may be hampering your development," King writes. "You may find, for instance, that a whole series of 'worst memories' in a particular session has a fear-of-rejection theme or a need-to-control theme." When it comes to women, we've all had embarrassing experiences. But if these incidents aren't properly handled in our subconscious, they can cause us to sabotage our own potential for success.

EMOTIONAL FREEDOM

One of King's main teachings is to stop being a victim to your subconscious, and instead learn to guide and instruct it.

One way to do this is by striving for what King calls emotional freedom. Stop identifying with "the emotional reactions of your subconscious," King writes. "When you say, 'I am angry,' you are identifying with the subconscious, and you may find it extremely difficult to get rid of the anger."

Instead, determine the purpose and origin of a new emotion as soon as it starts. Ask yourself, "Where did this emotion come from? Why am I feeling it right now?"

These and other questions allow you to discover the sources of your emotions. Even the act of self-examination itself can help you calm down. "The analysis itself tends to drain the emotion of its power because you are diverting the energy of the emotion to the conscious thinking process," King explains.

He also prescribes reprogramming as a technique to control your subconscious. "If you want to change the habitual thinking of the subconscious, you must consciously keep the desired pattern in the forefront of your mind until the subconscious has accepted it as a new habit." This is why affirmations, as silly as they seem sometimes, can directly improve your success with women.

THE CONSCIOUS MIND

To truly understand the conscious mind, it's necessary to understand the nature of will power. The only real ability you have on a conscious level is the power to direct your awareness and attention to a thought or experience. This is what's meant by "free will."

We can't make a woman like us, make the boss give us a raise, or make that 1974 Ford Pinto start in the morning. "What we can do, however, is to choose to decide how we are going to respond to our experience of life, what we are going to do from this moment forward and in any future moment to change either ourselves or the circumstances," King writes.

King defines determination as "the continuous, conscious directing of attention and awareness toward a given end for a purpose." And goals are achieved, he continues, "by continuously renewing the decisions or choices made to reach the given end, in spite of apparent obstacles and difficulties."

In other words, if one method does not work after repeated attempts, a determined person doesn't give up. "He tries another, and then another, until he finds one that does work, even if it means he has to change himself."

The difference, King concludes, between those with strong will and those with weak will is that the strong decide to continue, while the weak quit. It's important to remember this when the girl you've been talking to all night gives you a fake phone number, or you see a woman who just rejected your approach making out with some stranger. Failures and setbacks are fine. Deciding to quit is not.

GOALS AND PURPOSES

King makes a distinction between achieving goals and fulfilling a purpose that is key to your self-improvement journey.

The difference is that a purpose is "something that will give meaning to your whole life." A goal simply measures progress toward your purpose—like the concrete results you wrote down for your personal mission statement.

"Unlike a goal, a purpose is not something you reach but something you do," King writes. "Goals without purpose are empty of meaning, while having a purpose can give meaning to any goal."

Elsewhere in his book, King provides countless other tools for improving your mental and emotional states. By using your mind to improve your life, you can build the confidence that is an absolutely vital component to being successful with women.

As King suggests, "Look for the good in everything and, if you can't find any, figure out a way to put some in."

DAY

MISSION 1: Share Your Traits

Write down eight qualities you want someone to know about you. These might include individuality, humor, trustworthiness, intelligence, artistic talent, or whatever else makes you stand out.

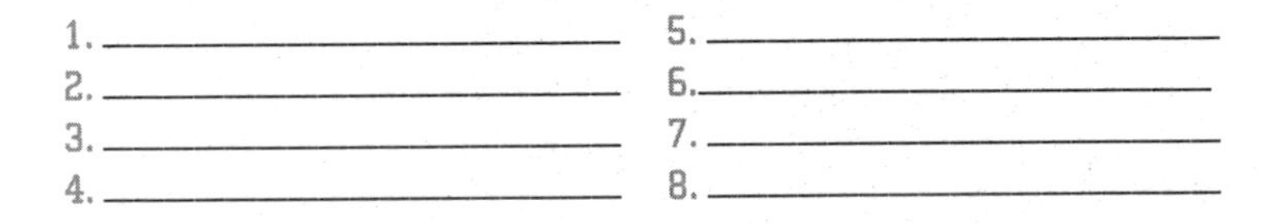

MISSION 2: Find Your Stories

Now you know what you want to convey. But how do you convey it?

Welcome to storytelling day.

Though most women tell guys that learning to listen is important, in the early stages of an interaction, learning to speak is more important. This is because it's your job to demonstrate you're worth spending the night talking to.

Your vehicle for doing this is your past. Rather than telling women your best qualities and most charming foibles, stories allow you to show them. They also prevent you from blitzing a woman you've just met with generic questions about where she's from and what she does for work. And they provide the opportunity not just to fascinate a group of people but to inspire them to share their own stories in return.

Your tasks today will lead you toward the generation and performance of the perfect story.

You may be lucky enough to be a great storyteller already—able to hold court at countless dinner parties with the tale of that one time you had to break into a drugstore in Cairo at three in the morning to get aspirin for your girlfriend.

Or perhaps you're less loquacious, unable to think of stories on the spot or to hold anyone's attention long enough to share them. I've heard hundreds of men claim that their lives aren't interesting and they have no stories to tell. This is just another limiting belief rearing its head. It doesn't matter how small a town you live in, how little you may have traveled, how normal your family is, or how old you are, you do have interesting stories to tell. You just have to find them.

So think of the memorable moments in your life, whether they're pivotal experiences that shaped who you are as a person or just funny, trivial anecdotes that you enjoy sharing. They might be:

- ironic and embarrassing, like the time you went to relationship counseling with your girlfriend, and the therapist asked her out afterward;
- adventurous and exciting, like the time you were scuba diving, your regulator broke, and a school of barracudas swarmed around you;
- sexy and awkward, like the time the married woman sitting next to you on the plane tried to have sex with you in the lavatory;
- naive and touching, like the time your hamster died and you thought it was sleeping—for seven days;
- small and poetic, like the time you were eating a burger and suddenly realized the meaning of life;
- dangerous and heroic, like the time you saved a girl from some guy who was threatening to beat her up outside a club in Rio;
- current and confusing, about something that happened only minutes ago, like a girl you don't know coming up and asking if you'll take her sister home;
- anything you want them to be—as long as they don't evoke negative emotions in listeners or hint at negative qualities about yourself such as misanthropy, stinginess, unhappiness, prejudice, anger, or perversion.

Now think back over your childhood, family life, school, work, travel, recreation, and dating experiences, from your earliest memory to what you did last night. Extract from those experiences eight personal stories. Then give them intriguing names (like "The Magical Hamburger Incident" or "The Festering Hamster Story") and write them down in the space below:

1. ______________________________
2. ______________________________
3. ______________________________
4. ______________________________
5. ______________________________
6. ______________________________
7. ______________________________
8. ______________________________

If you're having trouble coming up with eight stories, think back on recent conversations you've had with friends and family. Try to recall any anecdotes you told that elicited excitement, intrigue, or laughter.

If you're still having trouble, imagine that you have a chance to pitch a movie about yourself to film producers. What key stories from your life would you need to include to interest them?

If you're still stuck, call a parent, sibling, or friend, and ask them to share a few favorite memories about you.

MISSION 3: Select Your Stories

Your next task is to scan the qualities you listed in Mission 1. Then look over the stories you chose for Mission 2. Mark with an asterisk each story that displays one or more of your eight qualities. Note that an ideal story does not brag or overcompensate but displays both your strengths and your vulnerabilities in an honest, humble, humorous, and engaging manner.

Of the stories you've marked, choose the two that you find most compelling and entertaining. (If you haven't marked any stories with asterisks, it's time to think of more stories—or more qualities.) List your two top stories here:

1. ______________________________
2. ______________________________

These are the core stories you'll work on today.

MISSION 4: Prepare Your Stories

Grab a piece of paper, pull out your journal, or open a new file on your computer.

Write out each of the two stories in their entirety. Anything goes—as long as you don't fib, because it could come back to haunt you. Here are a few tips:

- *Have a strong beginning.* Your story needs to make a good first impression, and the best way to ensure that is to have a short, sharp, clear initial sentence. This can be a summary that flows naturally out of the conversation: "Oh, yeah, that's like the time I was forced to eat rancid shark in Iceland." It can take the form of a question that grabs the listener's interest: "Have you ever eaten rancid shark?" Or it can just be an intriguing hook: "The weirdest thing happened to me while I was in Iceland."
- *Have a good ending.* If the story takes a surprising twist at the end, reveals the answer to a mystery posed earlier, has a non-cheesy punch line, or wraps everything up into a neat lesson, this is ideal. Either way, make sure your last sentence leaves the listener with laughter, excitement, shock, admiration, disbelief, or any strong, positive emotion. You may also want to add a question at the end, to elicit responses or similar stories from your listeners.
- *Add intrigue.* Suspense occurs when a listener knows something is going to happen next but doesn't know either what it is or how it's going to happen. So make sure your audience is aware at all points where you're going with the story—or at least that you're going somewhere—but not how you're going to get there.
- *Include vivid detail.* Play back the experience in your mind as you write. Close your eyes if you have to. Remember sights, sounds, smells, and feelings. The richer the detail, the more involved the listeners will become.
- *Add humor.* Watch good stand-up comedians and you'll notice that between a set-up and a punch line, they squeeze in several additional jokes—plus a tagline after the punch line for an extra laugh. Find waypoints where you can add humor to your story. Useful devices include making fun of yourself, others, or human behavior;

comical exaggeration; references back to previous jokes; and saying the opposite of what people expect.

- *Add value.* When illustrating your positive traits, there's a right way to brag and a wrong way. The wrong way is to declare it in a sentence: "I just bought a new car." The right way is to share it as a casual detail that helps paint a picture: "So I was driving home, and I had to unroll the window because the new car smell was suffocating me."
- *Cut the fat.* When you're finished, reread your story. Make sure it's easy to follow and doesn't include unnecessary details and information. Mercilessly remove anything that doesn't contribute to the story. You may need to tell the story to a few people and make sure the pacing works.
- *Cut the neediness.* Make sure that the intent of the story is to entertain, amuse, or involve other people, not to sell yourself or your accomplishments. One way to prune validation seeking is to look at every instance of the words *I* or *me,* and see how many you can remove without detracting from the story.
- *Check the final length.* Your story should last no less than thirty seconds and no more than two minutes (that's roughly seventy-five to three hundred words on paper). If it's shorter, add more intrigue and humor. If it's longer, cut more fat.

Once you have both stories clearly written out, distill them to their major plot elements and make bullet points for each one. If, for example, you were describing *Star Wars,* the bullet points would be: Teenager living with aunt and uncle; buys two droids; discovers secret message; and so on. Unlike *Star Wars,* your stories should have only three to six bullet points.

Though you're going to practice reciting your entire story, all you need to memorize are the bullet points. This way, your delivery will seem less scripted, and you'll have more flexibility to expand and collapse the story, depending on your audience's interest level.

MISSION 5: Tell Your Stories

I have this theory about words. There's a thousand ways to say "Pass the salt." It could mean "Can I have some salt?" Or it could mean "I love you." It could mean "I'm very annoyed with you." Really, the list could go on and on. Words are little bombs, and they have a lot of energy inside them.
—CHRISTOPHER WALKEN

It's time to master the telling of your story.

The best way to captivate a listener is to be passionate. Be excited about your life, intense about your experiences, and believe in every word you say. Each time you repeat the story, it should seem like you're telling it for the first time—with all the confusion or excitement or wonder you felt when first experiencing it.

Review the vocal exercises from Day 3, then recite your two stories into your audio recorder. Make sure you speak loudly, slowly, clearly, and dynamically. To further hook listeners, stress key words and insert pauses to build suspense or humor. Experiment with emphasizing different words and pausing in unexpected places to change the rhythm of the story.

Once you're comfortable with your recitation, find a place in the middle of each story to insert an opportunity for listeners to interact. This will help keep their attention. Most interaction points will involve asking listeners if they relate to an experience, have an opinion on the experience, or can jog your memory with a fact.

For example, if you're telling a story that takes place at a Chuck E. Cheese's pizza parlor, your interaction point can simply be: "Have you ever been there? Okay, so you know what I'm talking about." If it takes place in an airport, you can ask: "It was kind of like that movie where Tom Hanks plays the guy stuck in an airport. What was it called?"

If you want to take your performance to the next level, practice casually pausing at the climax of the story to build suspense. You can take a sip of your drink, put a mint in your mouth, or, if you smoke, light a cigarette.

After you've made a successful recording of your stories, go back to the piece of paper or computer file where you originally wrote them and update them. Add any interaction points, pauses, or other embellishments you came up with while working on your delivery.

MISSION 6: Perform Your Stories

You've reached the final step in preparing your stories.

Stand in front of a mirror or set up a video camera to film yourself.

Watch yourself recite the story.

The key to a good performance is being expressive. Facial animation, eye movements, hand gestures, body language, and energy level can all tell a story as powerfully as the words themselves.

Experiment with accentuating different thoughts and emotions in the story with specific movements. Try changing your gestures or tone of voice when you're quoting other people. And feel free to use any props within arm's reach—a cell phone, a straw, or another person.

However, be careful not to overdo it. The smaller and more subtle your gestures and affectations are, the more credible they'll be. Don't get overly hyper or spastic, and make sure you have the attention and interest of the group at all times, allowing them to contribute when they want to. Don't blitz them with unrelated stories back to back; that could push you over the line from conversational expert to conversational terrorist.

There's one final element of the performance that you can't practice in front of a mirror: the unpredictable. As anyone who's been onstage will tell you, no matter how much preparation you've done, everything changes once the spotlight is shining on you.

So when you're talking to a group, don't worry about getting every gesture and phrase right. Just make sure you hit the bullet points. And if people ask questions, interrupt you, or suddenly start telling their own related story, don't get flustered. This is a good thing: It means they're paying attention.

If the conversation veers off course, don't insist on finishing your story unless your listeners ask what happened next. You can always keep the conclusion on tap for later in the evening to fill in an awkward conversational lull. Don't forget that the purpose of the story is not to get to the end, but to further display your magnetic personality.

On the other hand, don't tolerate rude behavior. Comedians deal with hecklers all the time. Have a few lines on tap for troubleshooting. A friend of mine, for example, jokes, "The show's over here," whenever someone gets distracted.

MISSION 7: Share Your Stories

Use your two stories—with interaction points—at least twice in conversation today. You don't have to tell the same person both stories; just make sure you use each story at least twice over the course of the day.

It doesn't matter whether you tell them to a woman you're interested in, a coworker, a friend, a parent, a stranger, a sibling, or a telemarketer, as long as you tell them.

Feel free to improvise. As you tell the stories, you may insert new details, jokes, and interaction points in the moment. After each successful recitation, return to your master story file and note anything you want to add, change, or remove to improve the telling.

If either of the stories doesn't hold your listeners' attention, replace it with another story from your list. If the new one doesn't work either, ask someone who was there at the time to give you feedback on your delivery or tell you his or her version of the events. If both stories get great reactions, start crafting new ones.

And congratulate yourself. Storytelling is one of civilization's oldest arts, and you're now officially part of that tradition.

DAY

MISSION 1: Get a Date Book

Turn to your Day 13 Briefing. Tear out the calendar page or make a photocopy. If you don't want to remove the page and don't have access to a photocopier, there's a copy available for you to print at www.stylelife.com/challenge.

MISSION 2: Promote Literacy

Head to a bookstore, preferably one with a café or sitting area. Bring the Stylelife calendar page, something to write with, and your journal, if you've been keeping one.

Get comfortable. You're going to perform the rest of today's tasks at the bookstore.

MISSION 3: Borrow Some Culture

Pick up a copy of a local events guide. This can be a free weekly newspaper, a magazine-style going-out guide, or a daily paper. You may also want to grab a local Zagat guide to restaurants or nightlife, or even a travel guidebook that includes local attractions. Since you won't be leaving the bookstore with these, you don't need to pay for them.

MISSION 4: Become Cosmopolitan

Pick up the current issue of *Cosmopolitan* magazine.

MISSION 5: Plan Your Nights

Sit somewhere comfortable in the store, like the café. Whip out your Stylelife calendar, and look through the listings, reviews, and recommendations in the paper or reference material you picked up.

Select an interesting event, restaurant, concert, gallery opening, reading, flea market, or other activity for each day of the week. Write the information for each event in the left-hand column of the calendar. The simpler and cheaper the activity, the better. Free is good too. Make sure it's something you are able to attend—not a concert that's sold out or a restaurant that's out of your price range.

In the larger column on the right side of the calendar, write one or two compelling sentences convincing someone why he or she should go to each event.

MISSION 6: Is That What They Really Think?

Read the issue of *Cosmopolitan* front to back.

First, note that women are just as desperate as men to get a date, keep a mate, and avoid rejection. Next, find an interesting topic of conversation inspired by an article, column, letter, or advertisement.

Once you choose a topic, comment on it to a woman seated nearby or wandering past. (If she's walking, speak to her while she's still coming toward you—if you see her back, you're generally too late.) Show her the story in the magazine, and tell her your reaction to it or ask a question about female behavior based on it.

If she responds favorably, then congratulate yourself. You've just generated your own spontaneous routine. If she doesn't, keep reading and find another interesting topic. Repeat with a different woman.

If she happens to ask why you have a copy of *Cosmopolitan,* tell her the truth: Someone recommended reading it to learn more about women.

There's no need to continue the conversation afterward. But if she's enjoying the interaction, feel free to proceed by using one of your openers, personal stories, or disqualifiers. Your mission is complete once you've talked about the magazine with three different women.

When you return home, add any *Cosmopolitan* routine you successfully used to the stories list you started yesterday.

DAY 13 BRIEFING

Sunday	
Monday	
Tuesday	
Wednesday	
Thursday	
Friday	
Saturday	

DAY

MISSION 1: Demonstrate Value

When you learned openers, one of the keys was to give a time constraint by saying you're going to leave shortly.

Your goal today is to be so cool and interesting that she doesn't want you to leave. The quickest way to reach this goal—the hook point—is to demonstrate value. After all, she has the possibility of meeting any number of guys that day. Why you?

For some women, just your having the confidence to approach might be enough to make you stand out from other men. For others, your sense of humor or your particular look may distinguish you. Perhaps you remind her of her first boyfriend, have a don't-give-a-shit attitude, or possess some other quality that excites her. But sometimes—especially with women who have a lot of options—you're going to have to do something a little extra.

One of the best and most efficient ways to make an impression is to teach her something about herself.

Your task is to turn to your Day 14 Briefing, read the note about using scripted material, then study the following routine and learn to give value to those you meet instead of taking value. Remember that, as with everything you've learned, there's no power in the routine itself. Your goal is simply to make her day or night better and more interesting than it was before she met you.

Once you've memorized the routine, move on to today's field assignment.

MISSION 2: Take Her Hand

Today you're going to add the rings routine to your growing repertoire.

Go any place where people gather—café, bar, park, museum, department store—and start a conversation using one of your new openers.

Afterward, as you learned on Day 11, pretend that you're about to leave. Then spontaneously notice the ring on her finger (or the lack thereof) and transition into the routine. Until you reach the hook point, and you're sure she's intrigued, continue to pretend as if you're about to leave at any moment.

If she's with friends, don't forget to include them in the conversation.

Her reaction to the rings routine doesn't matter. Whether she's fascinated or bored, you're doing this only to practice demonstrating value. Remember, these routines work best when performed in the spirit of curiosity and fun, not as a way to make an impression or get her to like you. As long as you're saying it and she's still standing there, you're completing the mission.

Feel free to continue the interaction if it's going well. If you don't know what to do after this routine, it's okay to politely make your exit. In the following week, you'll be given tools to continue the conversation, amplify the connection, and exchange numbers.

After you have practiced the rings routine on three separate women, your field mission for today is complete.

MISSION 3: What's Darwin Got to Do with It?

All this may seem like a lot of work.

After all, you're an amazing, unique individual. You've got your own life and family and friends. You're going places in the world. Why should you have to bend over backward just to meet the standards of some woman you barely even know?

The answer, my friend, is evolution.

Ultimately, whether you like it or not, in our species—and most species—men typically compete for women, and women choose men.

In your Day 14 Briefing, you'll find a book report by Stylelife coach Thomas Scott McKenzie on Matt Ridley's *The Red Queen*. Your assignment is to read the report and discover the evolutionary logic behind many of the things you've been doing this month. Keep in mind that cultural forces are at play as well in our behavior—though, of course, an evolutionary biologist would say that those forces are also shaped by natural selection.

DAY 14 BRIEFING

A NOTE ON THE MATERIAL INCLUDED IN THIS BOOK

One day I turned on the television and saw an episode of *CSI: Miami*. The plot was about a group of pickup artists using material that came word-for-word from my book *The Game*. It was the top-rated show in its time slot, reaching some fifty million viewers in fifty-five countries. Nonetheless, pickup artists around the world continued to use this exact same material, and I never heard a report of a single one getting caught because of the show.

So never underestimate people's capacity to forget the exact words they hear and where they came from.

But, for argument's sake, let's imagine a worst-case scenario: You run an opener, and the woman knows it came straight from the pages of this book.

No problem.

All you need is a contingency plan. And the premise of the plan is that you now both have something in common. You've both read the same book. So just drop the opener and exclaim, "No way. You know about the book. What do you think of it? I actually decided to test it out today—and on my first approach, I get busted!"

If the goal of the opener is to start a conversation, you're now officially in one, talking about one of the most interesting topics in the world: relationships.

There is no reason to fear any outcome you can imagine. Because if you can imagine it, you can prepare a contingency plan in case it happens.

In the bigger picture, remember that the language and wording don't matter nearly as much as the intent behind them. The shady friend opener works not because it's the shady friend opener but because it's a way to start an engaging conversation with a group of people without hitting on anyone. As long as you can always do that, you've got nothing to worry about if these techniques ever become widespread.

Knowledge won't change the fundamentals of how men and women are attracted to each other. And attraction, as you're about to read, has operated on the same principles since the dawn of man.

With that in mind, the following routine is just one example of demonstrating value. Feel free to study or use anything else during the Challenge that serves the same purpose—whether it be non-cheesy magic that doesn't involve cards, visualization games like the cube, personality assessment skills like handwriting analysis, or anything else that serves the end goal of being excellent.

THE RINGS ROUTINE

QUICK-START GUIDE

- Thumb = Poseidon, representing individuality, independence, and iconoclasm
- Index = Zeus, representing dominance, power, and energy
- Middle = Dionysus, representing irreverence, rebelliousness, and decadence
- Ring = Aphrodite, representing love, romance, and connection
- Pinky = Ares, representing conflict, assertiveness, and competitiveness
- No Ring = Hermes, representing friendliness, helpfulness, and adventurousness

PERFORMANCE SCRIPT

YOU: I have to ask before I run: Why did you choose to wear that ring on that particular finger?

HER: No particular reason.

YOU: Interesting. Do you always wear rings on that same finger?

HER: I guess. Most of the time.

YOU: The reason I'm asking is because I have a friend who's a spiritual type, and she just taught me that the finger you choose to wear your rings on actually says something about your personality. I don't know if I totally believe it, but she nailed my personality pretty accurately.

If she's not wearing any rings, use this alternative: "I have to ask before I run: I notice you're not wearing any rings. Do you usually wear rings?" Then continue with the paragraph above, but say, "She just taught me that wearing rings on certain fingers, or making the choice not to wear rings, actually says something about your personality."

> YOU: In ancient Greek culture, each one of the mounds at the top of the palm was represented by a different god. And people back then would wear a ring on the associated finger to honor that particular god.

Now go through her different fingers one by one. If it seems like you have time, save the finger her ring is on for last to build intrigue.

> YOU: For example, the thumb represents Poseidon, the god of the sea. And he was very independent. He was the only god who didn't live on Mount Olympus. And the thumb kind of stands apart from the other fingers. So people with thumb rings are generally independent thinkers who tend to do their own thing. They don't follow trends; they like to set their own.
>
> The index finger is represented by Zeus, the king of the gods. And it represents power and dominance. Just like when parents scold children, they always wave the index finger. So people with a ring on this finger generally have an inclination to take charge.

If she says that the finger her ring is on doesn't fit her personality, tell her that people sometimes choose those fingers because they're subconsciously working on cultivating that particular attribute or because they're attracted to people with that attribute.

> YOU: The middle finger is represented by Dionysus, the god of wine and partying. He was a very irreverent god. And he liked to free people from their inhibitions. So if you have a ring there, it means you tend to do whatever you want without depending too much on what others think. You can be an instigator sometimes. So it kind of makes sense that it's the finger people use to swear.
>
> Your ring finger is, of course, represented by Aphrodite. She was the goddess of love, and that's supposedly why we wear wedding rings on that finger. Interestingly, it's the only finger that has a vein

that goes straight to the heart without branching off. So when someone puts a ring on that finger, they're actually making a direct connection with your heart.

If she's comfortable enough with you to allow light touching, you can hold up her hand or touch her fingers as you do this. If she's shown more interest, you can even trace the line of her vein from her finger up her arm.

YOU: The pinky is represented by Ares, the god of war. That's why you see mobsters wearing pinky rings. It represents conflict. When people put the ring on themselves, back then it meant they were in conflict with themselves or had some inner turmoil. If it was given as a gift, that often meant there was an element of conflict or competitiveness with the giver beneath the surface.

If she's not wearing any rings, add the following:

YOU: People who didn't wear rings were aligned with Hermes, who was the messenger of the gods. He represented exotic travel and wealth, and loved the best of everything. But he wasn't greedy. He was known for his giving nature, and was the most helpful of the gods. He was also the most adventurous. So people with no rings tend to be open minded, and enjoy travel and being around others.

THE EVOLUTION OF SEXUAL PREFERENCE—A BOOK REPORT

By Thomas Scott McKenzie

In the book report on *Mastering Your Hidden Self*, we learned that everyone is shaped by his or her environment, experiences, beliefs, and expectations. In *The Red Queen: Sex and the Evolution of Human Nature* by Matt Ridley, we learn that we're also shaped by millions of years of evolution.

Understanding the evolutionary nature of attraction and mating, as well as the correlations in the animal kingdom, is essential in understanding our own sexual strategies.

According to Ridley, the most powerful tool we've evolved when it comes to meeting women is our mind: "Most evolutionary anthropologists now believe that big brains contributed to reproductive success either by enabling men to outwit and outscheme other men . . . or because big brains were originally used to court and seduce members of the other sex," he writes.

WHY MEN PREFER BEAUTIFUL WOMEN

Many men tend to think that women in their particular city or country are different and require a unique seduction strategy. Not only is this not true today, according to the experiences of tens of thousands of students, but it's not true evolutionarily as well. Wherever you go, the game largely remains the same.

"Until very recently the life of a European was essentially the same as that of an African," Ridley writes. He explains that both groups hunted meat and gathered plants, made tools from the same materials, utilized complex languages, and raised children in similar manners. Advances such as metalworking, agriculture, and written language, he continues, "arrived less than three hundred generations ago, far too recently to have left much imprint . . . There is, therefore, such a thing as universal human nature, common to all peoples."

He cites a study involving more than a thousand subjects in thirty-seven countries. The statistical evidence revealed that "men pay more attention to youth and beauty, women to wealth and status."

These universal principles of selection exist not because human beings around the world are shallow but because they want to bear as many offspring as possible and have their offspring survive. Thus, according to Ridley, the male obsession with beautiful women is not so much about form as it is about function: "Prettiness is an indicator of youth and health, which are indicators of fertility."

Even the saying that gentlemen prefer blondes, Ridley claims, goes back to a correlation between blondeness and youth.

WHY WOMEN PREFER HIGH-STATUS MEN

Men have it easier than women in the looks department. "In a survey of 200 tribal societies, two scientists confirmed that the handsomeness of a man depends on his skills and prowess rather than his appearance," Ridley writes.

Study after study has shown that women are attracted to personality, dominance, and status. "In a monogamous society, a woman often chooses a mate long before he has had a chance to become a 'chief,' and she must look for clues to his future potential rather than rely only on past achievements," Ridley writes. "Poise, self-assurance, optimism, efficiency, perseverance, courage, decisiveness, intelligence, ambition—these are the things that cause men to rise to the top of their professions. And not coincidentally, these are the things women find attractive."

In other words, if you exhibit the right traits for success, some women will take a chance on you even if you're currently unemployed.

One of those traits is body language. Ridley describes an experiment where scientists recorded an actor doing two fake interviews. "In one, he sat meekly in a chair near the door, with his head bowed, nodding at the interviewer, while in the other he was relaxed, leaning back and gesturing confidently," he writes. "When shown the videos, women found the more dominant actor more desirable as a date and more sexually attractive."

WHY POPULARITY MATTERS

Ridley points out that peacocks are among the few birds to gather together in groups for sexual selection. Scientists call this gathering a lek. "The characteristic of the lek is that one or a few males, usually those that display near its center, achieve the most matings. But the central position of a successful male is not the cause of his success so much as the consequence: Other males gather around him."

Elsewhere in the chapter, Ridley writes that in experiments with guppy fish, when a female is allowed to see two males—one courting a female, the other not—she later prefers the male who was with the female, even if the courted female is no longer present.

Female competitiveness and social proof—the idea that individuals emulate what they see others in their peer group doing—seem to be effective, even in the animal kingdom.

WHY WOMEN GET TO CHOOSE

The instinctual goal for female animals is to find a mate with the genetic makeup necessary to be a good provider or a good father. Male animals, on the other hand, have a goal of locating as many wives and mothers as possible.

The reason for these disparate goals is *investment*. The gender that invests the most in children (by carrying a fetus for months, for instance) is the one that has the least to gain from extra mating. On the other hand, the gender that invests the least in children has the most extra time to spend searching for additional mates.

These different goals lend a scientific authority to something every man who's entered a singles club immediately learns: Males compete for the attention of females.

Ridley continues, "The male's goal is seduction: He is trying to manipulate the female into falling for his charms, to get inside her head and steer her mind his way. The evolutionary pressure is on him to perfect displays that make her well disposed toward him and sexually aroused so that he can be certain of mating."

Ridley examines the mating habits that revolve around peacock tails, deer antlers, swallow tailfeathers, and the colors of butterflies and guppies. The bottom line is that "females choose; their choosiness is inherited; they prefer exaggerated ornaments; exaggerated ornaments are a burden to males. That much is now uncontroversial."

For many women, high heels, push-up bras, tight clothing, and waxed body hair are just part of being fashionable and attractive. If you want to be successful with women, you have to be willing to carry a similar burden. It may feel unnatural or uncomfortable sometimes, but wearing clothes that distinguish you from the herd conveys confidence, leadership and individuality (as long as the clothes aren't wearing you). As Ridley puts it, "There is no preference for the average."

WHY MEN PURSUE CASUAL SEX MORE THAN WOMEN

Ridley argues that our different attitudes toward sex are determined by consequences. Historically speaking, casual sex for a man was a fairly low-risk activity with a huge potential payoff: "a cheap addition of an extra child to his genetic legacy," as Ridley puts it. "Men who seized such opportunities certainly left behind more descendants than men who did not. Therefore, since we are by definition descended from prolific ancestors rather than barren ones, it is a fair bet that modern men possess a streak of sexual opportunism."

Conversely, women faced massive risks when it came to casual sex. In the generations before reliable birth control, a married woman could be left with a pregnancy and potential revenge from her husband. If she was unmarried, then she could be doomed to a life of spinsterhood. "These enormous risks were offset by no great reward. Her chances of conceiving were just as great if she remained faithful to one partner, and her chances of losing the child without a husband's help were greater. Therefore, women who accepted casual sex left fewer rather than more descendants, and modern women are likely to be equipped with suspicion of casual sex."

Ridley points to interesting studies that further support his theories on promiscuity, citing research estimating that 75 percent of gay men in San Francisco have had more than one hundred partners (25 percent have had more than one thousand), while in contrast most lesbians have had fewer than ten partners in their lifetime.

WHY MEN AND WOMEN CHEAT

One interesting conclusion suggested by Ridley's book is that human beings are naturally monogamous, but they're also naturally adulterous.

Though Ridley says that women are less inclined toward casual sex, that doesn't mean they aren't promiscuous. But their promiscuity often has a purpose. For examples, Ridley looks to the animal kingdom—specifically to the phenomenon of adultery among colonial birds.

Like many human beings, female colonial birds divide men into two different categories: lovers and providers. "When a female mates with an attractive male, he works less hard and she works harder at bringing up the young," Ridley writes. "It is as if he feels that he has done her a favor by providing superior genes and therefore expects her to repay him with harder work around the nest. This, of course, increases her incentive to find a mediocre but hardworking husband and cuckold him by having an affair with a superstud next door."

Ridley closes his discussion of this topic with a crude summary of the hunter-gatherer rules that he claims still exist deep in the minds of women: "It began with a woman who married the best unmarried hunter in the tribe and had an affair with the best married hunter, thus ensuring her children a rich supply of meat. It continues with a rich tycoon's wife bearing a baby that grows up to resemble her beefy bodyguard. Men are to be exploited as providers of parental care, wealth, and genes."

WHY MEN LIKE PORN MORE THAN WOMEN

One of Ridley's more interesting asides concerns studies on male and female arousal.

Men are generally aroused by visual images; hence the success of pornography and *Maxim.* But what is the equivalent of pornography for women? His answer: romance novels, which have hardly varied for decades.

What turns women on in romance novels, however, isn't their descriptions of dashing men or lurid sex. Sex in romance novels, he explains, "is described mainly through the heroine's emotional reaction to what is done to her—particularly the tactile things—and not to any detailed description of the man's body."

The point is that women are aroused through emotional reactions, and the key to these are words and touch. So to become a master seducer, you must become a master of language and the female body.

According to another study of heterosexual men and women, men are more aroused by group sex, while women are more aroused by heterosexual couples. Yet both heterosexual women and men are aroused by lesbian scenes, while neither is aroused by male homosexual scenes. So if you're one of those men who thinks that sending a woman a close-up naked picture of his abs or his genitalia is going to turn her on, think again.

WHY THE STYLELIFE CHALLENGE?

The Red Queen explains how our mating choices are the result of evolutionary and biological pressures exerted over thousands of years, providing scientific proof for the social improvement strategies discussed, such as dressing sharp, demonstrating value, raising social status, displaying personality, and projecting confidence.

Even the idea that your friends will give you a hard time as you improve is cited in this book as a normal evolutionary result of your success: Males want to destroy competitors, even the ones they secretly want to emulate.

And, finally, if you want to improve your confidence, Ridley says you're doing the right thing by going out and working to craft the perfect approach.

"We measure our own relative desirability from others' reactions to us," he writes. "Repeated rejection causes us to lower our sights; an unbroken string of successful seductions encourages us to aim a little higher."

MIDPOINT COACHING SESSION

Pull up a seat and let's talk about life.

Here's the secret of success: What you get out of something is equal to what you put into it.

Why am I telling you this now?

Well, as a wise website once told me, "People don't fail. They just stop trying."

The midway point is a dangerous time in most regimens, and I want to make sure you're not going to bail out on the brink of a breakthrough.

Maybe you're doing just fine and anxious to press forward. But if you're anything like the majority of past Challengers, you're beating yourself up mentally before and during the field assignments.

Over what? Why are you giving these strangers power over you?

They are walking sources of feedback—there to give you insight about yourself and teach you how to do better next time. They're not even judging you nearly as much as you're judging yourself over this.

If I'd gotten discouraged by all the rejection letters I'd received (not to mention by the incoherent prose of my first stories), I wouldn't be writing today.

But I learned from every paragraph, every mistake, every critique, every success.

So guess what?

This is a challenge. That means it's going to be challenging.

Not difficult, just challenging—to the bad habits that never worked for you in the first place.

You've been offered an olive branch to fix yourself.

Are you going to take it and run with it, or are you going to just stand there and hit yourself over the head with it?

Every single person I know who's dazzlingly successful with women worked hard to get where he is. Whether he admits it now or not, he's overcome amazing obstacles—the biggest of which has been himself.

All the frustrations (as well as the highs) you're experiencing as you complete these assignments, we've all experienced. And what separates the ones who succeed from the ones who don't is their commitment to themselves, to the game, and to getting in the field and playing their best.

One of the most frustrating things about the game is that it requires effort. No matter how much status you may have at work or in school, you don't have more status than that jaw-dropping woman who's dressed to kill and turning every head as she glides through the club. No one does. Not the rock star. Not the billionaire. She can have her pick of the litter. And she can pick you. But it's going to take commitment.

Every time you don't approach, every time you don't try, every time you give up on something, every time you just go through the motions, every time you talk yourself out of a new or uncomfortable experience, the only person who loses is you.

To quote Wayne Gretzky on hockey: "You miss one hundred percent of the shots you never take."

It's time to take that shot.

DAY

MISSION 1: Cold Reading

Today you're going to learn one of the quickest ways to distinguish yourself from other men. Using this technique, you can very quickly enter the minds of strangers and tell them things even their best friends may not know.

Your assignment: Turn to your Day 15 Briefing and read the primer on cold reading.

MISSION 2: See a Psychic (Optional)

Your mission: Go to a psychic and get a reading.

If you have a portable audio recorder, ask to record the session.

The only reason this is optional is because it costs money, typically from five to forty dollars—don't pay more. However, I strongly recommend it for all Challengers.

Most communities have a few storefronts, New Age bookstores, and street fairs where psychic readers can be found. If you don't know where to find a local palm reader, tarot reader, or other fortune-teller, check the guide to local events you read on bookstore day, or go to Google Maps (http://maps.google.com) and search "psychic mediums." If you still can't find one, as a last resort call the American Association of Professional Psychics at 1-800-815-8117 (or, internationally 1-561-207-2391) and buy a ten-minute reading by phone.

Warning: Though most psychics are trustworthy, some are not. So don't give out financial, credit card, or personal identifying information. In addition, you shouldn't pay more than the initially quoted fee; if they ask for money after the reading to warn you of an impending event, don't fall for it. Thank them for their time and leave.

MISSION 3: Rate Your Reading

The following task is for all Challengers, whether or not you've gone to a psychic. (If you haven't, for reasons of money or time, go to www.stylelife.com/challenge. Input the date, time, and location of your birth, and get your astrological chart. Read the information as if you're getting your fortune told.)

Spend a few minutes analyzing the information you received during your psychic session, based on the cold-reading article you read today. Ask yourself the following questions:

- Did you feel the reading was good or bad? Why?
- Did you feel the reader was performing a generic routine or genuinely connecting with you? Why?
- Did you feel the reader understood you less or more than some of your friends? Why?
- Do you believe the reader had extra sensory powers? Why or why not?
- Would you visit the reader again? Why or why not?

Take a moment to reflect on these answers and what they tell you about the characteristics of a good or bad cold reading. If there were any lines or phrases from the psychic reading that particularly resonated with you, write them down in the space below:

____

____

____

____

Consider using these lines and phrases when performing your own cold readings.

DAY 15 BRIEFING

THE SECRET ART OF COLD READING

By Neil Strauss, Don Diego Garcia,
and Thomas Scott McKenzie

Most Challengers fit a certain personality profile, known as the Explorer, and chances are that you're one of them. If so, the following analysis may apply to you:

Explorers acknowledge that they have a few personality flaws, but they're usually able to compensate for them with their ability to keep up appearances. This is because, beneath the surface, they have an incredible amount of personal potential just waiting to be tapped. They try to seek variety in their encounters and feel like a caged tiger when too many rules are forced on them.

Explorers have a tendency to be a little hard on themselves sometimes but find solace in positive encouragement. At the same time, they take pride in their independence and don't just blindly accept the opinions of others. That doesn't mean, however, that they don't have a part of them that wants—and perhaps even needs—to be liked by those around them.

As Explorers grow a little older, they develop more secrets. And though they continue to work on themselves and make progress, they sometimes look back and wonder if they've made all the right decisions in life. A few of their dreams remain achievable in the near future, while a couple of others are a bit fanciful.

If you found yourself nodding and agreeing at any point, you have just discovered the power of cold reading. In short, the art of cold reading is making a truism sound like a revelation. *Cold* refers to the fact that the person knows nothing about you. And *reading* refers to when your experiences, thoughts, desires, and future events are told to you as though they were lifted from the pages of a book.

And they were. The script above is based on a classic reading, which has been passed on through generations of fortune-tellers.

History

In 1948, psychologist B. R. Forer gave a personality test to his students. Regardless of how they answered, Forer gave everyone the exact same personality profile afterward. He then asked the students to evaluate the accuracy of the profile. A score of 5 meant that the recipient felt the profile was excellent.

The class average turned out to be 4.26. So all these unique, individual human beings were told the exact same thing, yet they felt the words fit them almost entirely accurately. The conclusion: People tend to accept vague and general personality descriptions as being completely relevant to themselves. Furthermore, people usually accept claims about themselves in proportion to their desire that the claims be accurate.

These principles help explain why palm readers make a living, why people devour horoscopes in the paper every day, and why psychic hotlines exist.

Cold Reading and Attraction

If everyone's favorite subject is himself or herself, imagine the excitement they must feel when they meet a stranger who seems to know them almost as well as they know themselves.

So it's no surprise that cold reading occupies a central place in the art of attraction. Here are just a few of its uses:

- *The cold-reading opener:* Making an intelligent observation or sharing an intuition about a woman can be an effective way to spark her curiosity, and prompt her to stop and talk to you. Phrases like "I have an intuition that . . . ," "Something tells me that . . . ," or "I just noticed that . . ." are good ways to preface your observation.
- *The cold-reading hook:* Sometimes it's necessary in an interaction to demonstrate that you stand out from the tools who usually come on to her. If you say something incredibly insightful and perceptive about her early in the conversation, she may begin to realize that she's met someone rare and special.
- *The cold-reading amplifier:* Yesterday you learned the rings routine, one of many tests, games, and demos at your disposal for showing higher value. A knowledge of cold reading is essential to turn these

demonstrations from mildly amusing ways of killing time to emotionally connecting experiences.

Ethics

Keep it positive.

Never predict anything negative in the future or anything that will cause harm. When pointing out a personality flaw, even if it's accurate, present it in a reassuring way.

Don't tell her, "You're really insecure." Instead, say, "You may not be the most confident person in the room, but deep down you know your own value."

Never use cold-reading scripts in a callous, manipulative way—especially as a scam to con women into believing that they share an intense, metaphysical connection with you. Instead, use cold reading as a legitimate conversation starter, connection builder, or way to demonstrate your unique knowledge of human behavior.

Finally, cold reading is a secret art that's traditionally passed from teacher to student. Do not use the term *cold reading* with the women and groups you approach, and do not share this information.

Vehicles, Props, and Systems

A cold reading can consist of just a line or two of insight about the person you're talking to, or it can fuel a half-hour-long demonstration of value.

A prop, classification system, or something specific to anchor your cold reading will give you the credibility, authority, and pretext you need to make your reading as long as you want. In general, save readings that last more than a few minutes for quiet environments and one-on-one moments after the hook point.

Any number of props exist to give authority to cold readings. These range from well-known tools like tarot cards, rune stones, and the book of I Ching to more esoteric forms of divination like scrying (crystals) and cubomancy (dice). If you don't want to carry around these items, there are many systems that require nothing but knowledge, including palm reading, numerology, astrology, and the rings routine you learned yesterday.

When meeting women in bars and clubs, you can also springboard into

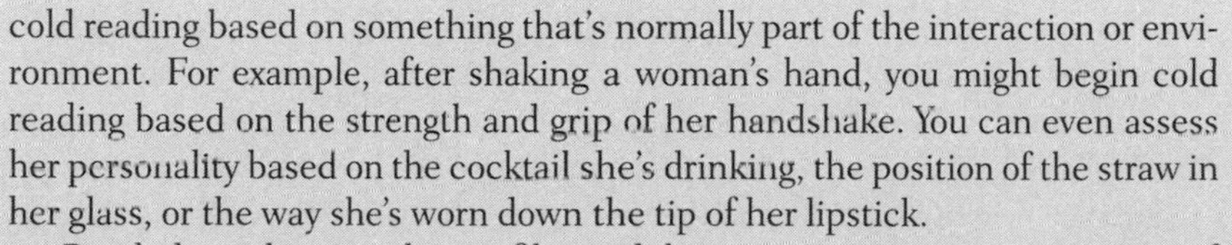

cold reading based on something that's normally part of the interaction or environment. For example, after shaking a woman's hand, you might begin cold reading based on the strength and grip of her handshake. You can even assess her personality based on the cocktail she's drinking, the position of the straw in her glass, or the way she's worn down the tip of her lipstick.

Psychological personality profiles and their accompanying jargon are one of the best ways to give your reading an increased air of authority and expertise. One such system is the social-styles model, which places people into one of four categories depending on their assertiveness and responsiveness. Here are the broad strokes of how it works:

To evaluate her assertiveness, ask if she's the kind of person who asks her friends what they want to do when they go out or tells them what the plan is. To figure out her responsiveness, ask if she's the kind of person who tells people when she's upset or keeps it to herself.

Based on her answers, you can create a cold reading based on the social-styles personality type she falls into:

- If she makes plans by asking for opinions and keeps her emotions to herself, then she's an *analytical.*
- If she makes plans by telling her friends what she's doing and keeps her emotions to herself, she's a *driver.*
- If she makes plans by asking and shares her emotions, she's an *amiable.*
- If she makes plans by telling and shares her emotions, she's an *expressive.*

Each of these personality types is associated with further behavioral traits, which can be researched online. Other systems worth looking into include the Enneagram and the Myers-Briggs Type Indicator.

The Cold Reading Code

Remember the Golden Rule—always tell the subject what he/she wants to hear!

—RAY HYMAN, "GUIDE TO COLD READING"

There are standard guidelines and principles that fuel every cold reading. Many have been around for centuries. Here are a few of them:

CONDITIONALITY

The key principle of cold reading is to never state an impression as a definite fact. It's far safer—and more accurate—to use conditional words and general terms.

If you say, "You are shy," your listener can always respond, "No I'm not."

But if you say, "You can be shy at times," this is a lot harder to deny.

When you use conditional words, every line you say during a cold reading becomes practically irrefutable. Here are a few examples of words and phrases to preface your insights with when developing your own cold-reading material: *a part of you, at times, every now and then, somewhat, generally, now and again, occasionally, once in a while, frequently, tendency,* and *sometimes.*

If you have some idea of your partner's disposition or the universality of your statement, you can use words and phrases with a narrower range of interpretation—like *usually, often, rarely, seldom, many, hardly, normally, regularly,* and *almost never.*

Unless you know for certain that your information is accurate, avoid absolute words and phrases such as *always, completely, every, all the time, none of the time, entirely,* and *never.*

FALSE SPECIFICITY

Though you want to avoid absolutes, this doesn't mean you can't punch up your reading with phrases that imply specificity.

One way to accomplish this is to use transition words like *because,* which imply causality even when a link doesn't exist.

Another way to make your reading sound specific is to affirm the listener's individuality by showing how her traits contrast with the norm. This can be accomplished by using a sentence structure like: "Though many people __________, you tend to __________."

CONFIDENCE

Act as if you're certain that everything you say is true. Even when you make a mistake or claim something that's not entirely accurate, if you say it with authority, most people will still believe it. On the other hand, doubt in your voice will create doubt in the listener's mind—even if what you're saying is true.

APPROVAL

People are more likely to agree with a positive statement about themselves, even when it's not true. Conversely, they're less likely to agree with a negative statement, even if it's accurate.

Welding these two principles together helps create one of the most powerful and beneficial things you can do when talking to a woman: to recast what she or others believe to be her negative traits into more positive ones.

If she's shy, for example, tell her, "Though some people think of you as shy, the truth is that you just take a while to get comfortable around new people."

Or if you're talking to a beautiful woman who's a little icy, you can tell her, "Some people think you're stuck up, but that's not true. You're just uncomfortable sometimes, and because of the way you look, people mistake your shyness for meanness."

AFFIRMATION

This is a simple technique that will make a big difference in a woman's judgment of your accuracy. Whenever you can, pause and get her either to agree explicitly with what you're saying or just respond with words like *yes* and *right*. The more yes responses she gives, the more her subconscious mind will accept you as an authority.

OPPOSITES

Some of the most powerful cold reads make a statement that contrasts two opposite qualities. For example: "At times you can be outgoing and social, while at other times you're more comfortable keeping to yourself."

This may look completely meaningless on paper, but try it out. When said with authority and understanding, it can seem incredibly insightful.

An additional technique you can use when delivering statements that contain opposites is the two-hand comparison: Lift one hand and indicate to it when you recite the first personality type, then lift and present your other hand when you describe the second type. Typically, her eyes or nose will point to the hand she has more affinity with as she considers each one.

OBSERVATION

When cold reading, it's important to be acutely aware of her reactions and facial expressions. Check to see whether her body language is affirming what you say (associative) or denying it (dissociative).

For example, without even realizing it, many people nod their head up and down while you're saying something they agree with, and shake it from side to side when they disagree. They may blush when you say one thing, and frown when you say another.

Below are examples of encouraging and discouraging cues to look for:

Associative Responses	*Dissociative Responses*
Head nodding up and down	Head shaking side to side
Eyebrows raising	Eyebrows lowering
Eyes widening	Eyes squinting
Smiling	Frowning
Body turning toward you	Body turning away from you
Animated expression	Blank expression
Arms open	Arms crossed

LISTENING

Often, people will start talking when you're cold reading them. Be quiet and listen, nodding and smiling as if these are things you already knew about them. They'll usually offer all the information you need to craft an extremely precise reading.

ADDITIONAL CLUES

When you're speaking face-to-face, you don't have to stick to scripted lines. Your eyes and ears can pick up a wealth of clues to help refine your reading. Pay focused attention to what she says, what she does, and the people with whom she chooses to surround herself.

A woman's age, ethnicity, speaking voice, style of dress, accessories, hairstyle, and jewelry are the most obvious signs of who she is. Look at her fingernails to see if they're clean or dirty, short or long, natural or painted. Notice the way she speaks, holds herself, and gestures. Does she do it with confidence or insecurity, and how does this relate to the way she looks?

Even where she's from—especially if it's a town associated with a particular university, company, or occupation—can give you extraordinary information. The more you notice, the more specific and accurate your cold reading will be.

TROUBLESHOOTING

It may just happen that as you're delivering a line, you see the woman you're talking to shaking her head negatively and folding her arms. If this happens, you need to recover. To do so, just stick to the rules: Assert your confidence and return to your conditional words. You can turn it all around through the power of just one word: *but*.

For example, if you're telling her, "You tend to be critical of yourself sometimes," and she starts to disagree, don't get flustered. Just continue speaking as if she's interrupting before hearing the complete thought: "But most of the time, you're more accepting of yourself. And this is what makes you stand out from others around you."

Be forewarned that there's one type of person you may come across who is invulnerable to cold reading. This person is what's known as a "polarity responder." Whatever you tell polarity responders about themselves, they're going to disagree. They may even get upset or angry that you claim to know anything about them.

For example, tell a polarity responder that she tends to be shy, and she'll probably respond, "I'm actually very confident." If you then simply repeat that she's confident, she'll say something like "Not always." Why? Because polarity responders just don't want to be defined. They derive their identity through their unique, uncompromising, often argumentative individuality.

Trying to cold read this kind of person is like trying to grip an eel. Eventually, you have to use a net to catch the eel. And that's exactly what you're going to do. Just smile and ask: "So you're the type of person who doesn't like to be pigeonholed?"

There is literally no way she can answer this without agreeing with you. As you watch her forehead crease and the befuddlement begin, just laugh with her, tell her you're joking, and quickly move on to another subject that doesn't involve cold reading. If her personality is really unpleasant, extricate yourself politely with your all-purpose farewell: "Nice meeting you."

Amazing Yourself

There's a next level to this talent.

Imagine walking up to a complete stranger and saying, "Out of curiosity, were you raised in a military family? . . . Yeah, I thought so . . . And you're probably the oldest sister too . . . I knew it!"

As you practice cold reading, you'll develop a strong intuition for people. Eventually, you'll find yourself going far beyond the principles described here, and you'll actually be able to guess with decent accuracy whether someone is an oldest or youngest child; what she does for work; what type of environment she was raised in; and any number of specific facts about her.

And if you happen to be wrong, you'll have the cold-reading skills to explain what led you to your conclusion in a way she'll ultimately agree with.

Sound impossible? Well, you'll learn more about how to do this when you study calibration on Day 28.

DAY

MISSION 1: The Missing Link

You have only one assignment today.

It's a piece that you've most likely not read about, heard about, or even imagined was part of the game. It's also subtle and will require the social and cold-reading skills you've learned thus far.

It's a piece that differentiates those who fail at the game from those who succeed—even though both may be saying the exact same thing.

It's a piece that will keep you from accidentally losing yourself in the effort to improve yourself.

It's also simple and basic. And it's one of the biggest lessons I've learned since writing *The Game*.

When I first started teaching workshops, I noticed that I could tell just by looking at a student whether or not he was going to get good reactions from women. And it had nothing to do with what he was wearing, what he looked like, or what he said. It was something intangible. A certain energy he gave off.

That's when I realized that everyone I'd met in the game, students and teachers alike, was overlooking something. But I didn't realize what that something was until months later.

Here's what happened: I had a student who'd been studying seduction for years. He's a sweet, good-hearted guy who knows every routine (even listens to them incessantly on his iPod) and goes out to meet women nearly every night. Yet he's still a virgin.

So he decided to fly to Los Angeles for a one-on-one session. He wanted me to examine him and find his Achilles' heel. I eventually found it, and it turned out to be such an epiphany that it changes the game of anyone who understands it.

Here is the key distinction:

The guy who fails at the game is the one who goes out looking for women to make him feel good about himself.

The guy who succeeds at the game is the one who goes out and makes other people feel good about themselves.

This first type of guy is someone no one wants to be around. He is needy, insecure, and reaction seeking. He will suck your energy dry in his quest for validation and approval.

This second type of guy is easy to be with. He radiates charisma and positive energy. Women enjoy his company, as do their friends, and they want him around all the time. They trust him, feel comfortable with him, and end up at his house at five o'clock in the morning wondering where all that time went.

Both guys do and say the exact same things, but they get very different reactions from women solely because of the intentions they're communicating.

Wait a minute, you may be thinking, what about disqualification? Doesn't it seem to contradict the idea of making people feel good about themselves?

Think again.

When you give a generic compliment to a woman who's often hit on, she'll usually ignore the remark—or assume you're saying it because you want to sleep with her. So instead you tease her, show her you're unaffected by her beauty, and demonstrate that you're out of her league. When she works to win *you* over, and you ultimately reward her with your approval, she will leave that night or the next morning feeling good about herself—like something special has happened and she's connected with somebody who appreciates her for who she really is.

In short, a teasing disqualification will buy you the credibility you need to sincerely compliment her later.

So today we're going to let go of our need for approval and we're going to make people feel good about themselves. Don't go to bars looking for approachable groups or cafés looking for lone women. Just go about your daily life. But three times during the day, go out of your way to make someone feel good about himself or herself.

That is your mission.

This might include telling a parent how much you appreciate him or her; making an awkward guest at a party feel wanted and included; telling a person who just blew a lot of money on a new outfit or haircut that it looks good; giving a homeless person eye contact, smiling, and handing over five dollars; or asking someone in a rush if they'd like to cut ahead of you in a checkout line.

Be sure to look for what people need when you do this exercise. Don't just give random compliments. And don't be concerned with whether you're raising or lowering your relative status. For example, if you see someone getting out of a new yellow Lamborghini with the dealer plates still on, instead of thinking he's an asshole and a show-off, consider that he spent a lot of money because he wants your approval. So give it to him: "Hey man, cool car. I'm jealous."

Of the three people you make feel good about themselves today, only one interaction can occur on the phone. And at least one of the people you interact with must be a stranger.

The goal is to stop worrying about what other people think of you, and start developing an instinct for what they need to feel good about themselves and their choices. You'll be amazed by the results.

After spending the weekend in L.A. and discussing these ideas, the student with the former Achilles' heel sent me the following email: "The other night, it was my twenty-sixth birthday. I was chatting up a four-set using the positive ideas we'd discussed, and one of them started groping me. Next thing you know, hardcore tongue-down makeout. First time ever!"

So get out of your head and start mastering the most intelligent and evolved emotion there is: empathy.

DAY

MISSION 1: Make No Mistake

We've covered a lot of ground in the last sixteen days.

So let's pause and make sure you're up to speed.

Welcome to review day.

Your first task is to go over the Day 17 Briefing, which covers the eleven most common mistakes guys make when opening.

Make sure you're no longer doing any of them.

MISSION 2: Check Your Core Competency

Look over the previous eight days and review each mission.

Make a list of the skills you don't feel you've mastered yet.

Your assignment is to redo each and every task you don't feel competent in.

At this point, you should be able to walk up to a woman or group, deliver an opener successfully, and transition smoothly into a value demonstration such as the rings routine. In addition, make sure you haven't slacked in your attention to your body language, speech, and appearance.

MISSION 3: Return of the Rings

Your final review mission is to go out, approach a woman or group, and perform the rings routine again.

Take your time with the delivery and incorporate the cold-reading information you learned on Day 15. Try to get a feel for the personality and self-image of the person you're talking to. Add at least one of the cold-reading scripts you heard or read on Day 15 as well as an original line based on your own assessment of the person. Notice her responses to the material.

Your mission is complete once you've successfully demonstrated the rings routine, with additional cold reading, for two different women or groups.

DAY 17 BRIEFING
THE ELEVEN COMMANDMENTS

1. Don't wait to approach her until she's alone. Even if she likes you, her friends will soon drag her away.

2. Don't stare at her for more than three seconds before approaching. Hesitate, and you'll either creep her out or psych yourself out.

3. Don't be afraid to approach just because there are men in the group. Chances are she's with family, friends, or coworkers, not a love interest.

4. Don't open a conversation by apologizing. Phrases like "Excuse me," "Pardon me," and "I'm sorry" make you sound like a beggar.

5. Don't hit on her or give her a generic compliment. Instead, start a conversation with an entertaining anecdote or question, such as asking the group to suggest names for a three-legged cat or a store that sells 1970s memorabilia. Everyone loves to give an opinion.

6. Don't buy her a drink. You shouldn't have to pay for her attention.

7. Don't touch or grab her right away. If she touches you, say, with a smile, "Hey now, hands off the merchandise."

8. Don't lean in or hover over her. Stand up straight and, if the music's too loud or she's seated, simply speak up.

9. Don't initially ask what her name is, what she does for a living, or where she's from. She's bored of talking about the same things with every new guy she meets.

10. Don't focus all your attention on her when she's with other people. If you win over her friends, you'll win her.

11. Don't be afraid to violate any of these guidelines once you understand them and why they exist.

DAY

MISSION 1: Sewing a Conversation

So far, you've learned a sequence of things to say and do when meeting a woman. It's now time to find out how to stitch it all together and leave her wanting more.

Your mission: Read the Day 18 Briefing on the four secrets of compelling conversation before moving on to the next assignment.

MISSION 2: Think Fast

In improvisational comedy, there's an exercise called the herald. To begin a herald, an actor asks the audience to suggest a word. As soon as an audience member offers one, the actor tells a true story from his life based on that random suggestion.

The story doesn't literally have to be about the word: It can simply be something that the word suggests or reminds him of. For example, if the audience member says "clown," the monologist can share a memory about his first time visiting the circus, about acting like a class clown in high school, or even about something that made him extremely happy or sad one day.

Afterward, the rest of the actors onstage improvise scenes based on his story; words or details in his story; or ideas his story suggests.

Your assignment is to try something similar at home: Practice spontaneously sharing true stories from your life based on one-word suggestions.

There are two ways to do this:

- Get together with a friend and take turns giving each other random words to elicit stories. It's important to start telling the story immediately, without waiting longer than ten seconds.

- Go to www.stylelife.com/challenge. I've created a random word generator there. Base a story on the random word presented by the generator. Make sure you recite the story out loud.

Practice this exercise until you feel confident spinning a story on the spot, with a definite beginning, middle, and end, based on an arbitrary word. If you're having trouble, reread your storytelling tips from Day 12.

The goal is to develop the ability to continue a conversation effortlessly using whatever material the woman you're speaking to gives you. Every concrete word she says is a hook you can choose to pull and stretch into a story or further conversation.

MISSION 3: Multiple Thread Mission

Today's field assignment is to practice creating the open loops and multiple threads you read about in today's briefing. You're going to do this by going out and delivering an opener. This time, though, before you finish discussing the opener, start another thread.

For example, if you're delivering the shady friend opener and you want to open a new thread, all you need to do is make a spontaneous observation or excited comment. You can interject, "By the way, I have to ask, why are you wearing a ring on that particular finger?" Or you can say, "Before we get to that, you'll never believe what just happened on the way here."

Possible threads include: one of the stories you developed on Day 12, another opener, an observation about her or something in the environment, a spontaneous story inspired by something she said, or a value demonstration like the rings routine or the social styles personality assessment.

Don't worry if this feels awkward or initially makes it seem like you have attention deficit disorder. Just approach and open, and you'll find that starting another thread will come easily once you have your mind set on it.

Your assignment is complete once you've approached two groups and successfully interrupted each opener with a second thread.

Note that creating open loops during your opener isn't a necessary part of most walk-ups. However, it is important to practice doing it today.

DAY 18 BRIEFING

THE FOUR SECRETS OF COMPELLING CONVERSATION: LOOPS, HOOKS, THREADS, AND . . .

Loops

The story collection *1001 Arabian Nights* begins with King Shahryar's discovery that his wife has been unfaithful. He kills her and declares that he can no longer trust any woman. From then on, he marries a different woman each day, spends the night with her, then executes her in the morning before she can cheat on him.

This reign of paranoid terror continues until one day he marries his match. Her name is Scheherazade. She knows that the king is planning to kill her in the morning. So on her first night with him, she starts telling a story. But just as the story reaches its climax, dawn breaks, and she stops at a cliffhanger and promises to continue the story the next night.

Curious to hear how the story ends, the king decides not to kill her that morning. And so it continues for night after night of cliffhangers, until Scheherazade has borne the king three sons, convinced him that she is faithful, and won his heart.

The principle Scheherazade employs is one known in the psychological field of neurolinguistic programming as open loops.

Simply put, creating an open loop means leaving a story or thought unfinished. This is the reason that TV series like *Lost* are so successful. Every week, these shows add more and more open loops to the plot, leaving viewers anxious for resolution on dozens of different mysteries.

When I was first learning seduction, if I wanted to get a woman's phone number or email address, I'd begin a value demonstration like the rings routine. But before I could finish it, I'd say I have to go meet friends or I'd have a friend pull me away. This way, if she wanted to hear what the rings on her fingers meant, she'd have to talk to me again.

Hooks

When talking with a woman you've just met, whenever she speaks, imagine the sentence or comment as a long horizontal string. Then imagine that there's a hook hanging down from each major word in that sentence. You have the option of pulling on any one of those hooks to start a new conversational thread.

Even a mundane line like "I've been working as a paralegal for six months" offers multiple hooks you could pull. You might tell any law-related stories you know; find out what she was doing before getting the job; ask about the office where she works; ask what exactly a paralegal does; tell her a story about one of your worst or best jobs; ask her opinion on a recent trial in the news; discuss the challenges of surviving law school; find out if she's new in town; or tell her to quit her job because you can get her a position as chief counsel in your little brother's lawn mowing business.

Even though she's hardly given you any information, she's created an endless array of hooks for you to pull on. And you can turn any of them into stories or humorous disqualifiers. To be a winning conversationalist, you generally want to grab the least obvious but most interesting hook.

Hooks also work in reverse. Instead of asking a woman questions, you can leave dangling hooks in your own conversation, selectively leaving out specific information in a way that compels her to ask you about your life. For example, if you say, "Back where I'm from, we don't do that kind of thing," she's bound to ask where you're from. Saying "Well, that may be true, unless you're in my line of work" will lead her to ask what you do. And now she seems to be chasing you.

Threads

Simply put, a thread is a single topic of conversation. For example, if you approach a group of women and deliver the shady friend opener, the thread would be the topic of jealous girlfriends. After ten minutes, though, that thread will start to wear thin. And if, in an act of desperation, you attempt to prolong the conversation by asking, "Well, what about girls who are friends with their ex-boyfriends?" it will seem as if you have nothing else to talk about.

The way to prevent this is to avoid focusing a conversation on only one topic and beating it into the ground. Instead, weave in several topics or stories at once, so that, like Scheherazade, you leave your audience captivated and wanting more. Juggling multiple open loops in a conversation will create the impression that you and the person you met have a lot to talk about.

Here's an example of creating a second thread during an opener, based on material generated during the Challenge by one of your colleagues.

> YOU: Hey, maybe you can help us settle a debate. Was there a fireman in the Village People?
>
> HER: I don't know. There was a construction worker and some leather guy.
>
> YOU: Yeah, there were five of them. And we can only figure out like, four: There's a cop, an Indian . . . By the way, really quickly, before I get to that, I just noticed your bracelet. My sister bought herself one just like it for her birthday.
>
> HER: Thanks. This was a present too, actually.
>
> YOU: Yeah, I always find it funny when people buy themselves presents for their birthdays. I mean, that doesn't count. Like one time, for my twentieth birthday . . .

Rather than talking for ten minutes about the Village People, you've started a second conversation in the middle of the opener. So when you're done talking about bracelets and presents, you can avoid an awkward silence by returning to the open loop about the Village People.

The most natural way to add a new thread to a conversation is by spontaneously noticing something new and getting more excited about it than what you were originally talking about.

This may sound artificial, but it happens all the time. Perhaps you're talking to a friend about a woman you met at the bank, but as soon as you name the bank, he interrupts to mention that he has a massive crush on a teller there. Or you're in the middle of a story, an ex-girlfriend suddenly walks past, and you pause to point her out to your friend.

Simply being aware of how to use loops, hooks, and threads can enhance your ability to make a deeper and more exciting connection with someone you've just met. They help create instant rapport, prevent potentially fatal pauses in conversation, and leave her with the impression that you two have a lot to talk about.

The Fourth Secret

Don't forget the moral of *1001 Arabian Nights*. As a species, we thrive on stories and suspense. So experiment with leaving routines unfinished, stories cut off at cliffhangers, and unresolved questions lingering in her head.

It can be as simple as saying, "There are three things I'm attracted to in other people, but I can't tell you the third thing yet because I don't know you well enough."

You can always choose to close the loop later in the conversation, during a future phone call or meeting, or never. If you leave her wanting more, you'll leave her wanting to see you again.

Finally you may be wondering about the fourth secret to compelling conversation? And I'd like to share that with you. Sometime.

DAY

MISSION 1: Fill Up Your Calendar

Pull out your Stylelife calendar—or print or copy a new one.

Fill in activities on the calendar—as well as selling points and reasons to go to each event—for today and the following six days. The items can be anything from restaurants to concerts to parties to roadside attractions to the psychic you went to on Day 15.

Familiarize yourself with the activities, the dates you listed them on, and the reasons for going.

MISSION 2: Seeding

You're now ready to begin the process of comfortably getting a woman's phone number.

Your first step: Turn to the Day 19 Briefing and read the short article on seeding.

MISSION 3: Seeding Mission

Seed three conversations today with an event from your calendar.

Two of these conversations can be with people you already know. However, at least one must be with a woman you've approached using one of your openers.

It isn't necessary to invite her to the event at the end of the conversation. The goal of today's exercise is not to get a phone number or a date (although if that does happen, great). The goal is simply to practice sprinkling a casual conversation with the seed for a future meeting.

DAY 19 BRIEFING

WHAT IS SEEDING?

Asking for a phone number can be one of the most difficult parts of an interaction with a woman you've just met. If she declines to give you her number, or instead asks for yours because she claims she doesn't give her phone number to guys, then all your previous efforts to build a connection with her have been in vain.

Even if she likes you, she may still refuse to give you her phone number the first time you ask. This is what's called an automatic or autopilot response: After experiencing repeated clumsy pickup attempts, many women have lines they use, almost by instinct, to politely decline requests for their phone number.

So what's the solution?

Don't ask for the phone number at all.

Today and tomorrow, you'll learn the two keys to exchanging phone numbers without asking.

The first key is seeding, a technique in which you mention a tempting event but do not immediately invite the woman to attend. For example, casually mention a party you're going to, talk about how cool it's going to be, and move on to other topics. Then, later in the interaction, before you're about to leave, decide to invite her to come along.

At some point in conversation with a woman I've met, I may mention my favorite local chef:

"You remember the Soup Nazi episode of *Seinfeld*? Well, this guy is the Sushi Nazi. His menu is only two words, 'Trust me,' and he just serves you what he wants. If you don't eat it in one bite, he'll stop serving you. If you dip it in soy sauce when he asks you not to, he'll cut you off. And if you dare ask for Americanized sushi, like a California roll, he'll chew you out and kick you out. But it's worth it, because the sushi literally melts in your mouth. The guy is an artist. He never smiles. He's just driven by some compulsion to make the best sushi in the world."

After I tell the story, I may even mention that I'm going there with friends on Thursday night. The obvious and expected next step would be to ask her immediately to join us. But because it's so obvious, I don't do it. I move on to other subjects and let her wonder why she wasn't invited. Only at the last min-

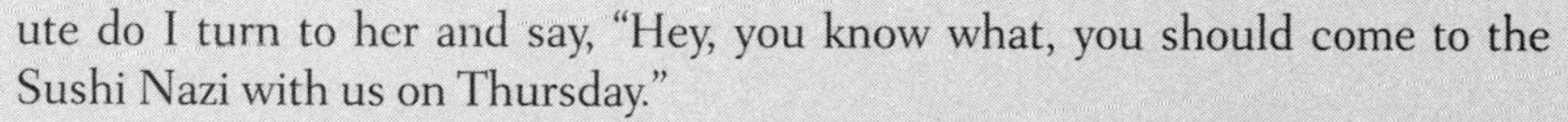

ute do I turn to her and say, "Hey, you know what, you should come to the Sushi Nazi with us on Thursday."

Sure, maybe I could have invited her when I first mentioned the restaurant; maybe she would even have said yes. But the point of the game is to eliminate the word *maybe* as much as possible from interactions with women.

Seeding helps to increase the odds of her saying yes, by avoiding the kind of pressure she might feel when confronted with a sudden invitation—pressure that often triggers a negative autopilot response. Mentioning the event, and then allowing her time to think about whether she wants to go before you get around to inviting her, gives her a chance to come to an affirmative decision on her own. Especially if you've continued to display more great personality, value, and non-neediness along the way. In addition, as you learned on disqualification day, not inviting her when you first mention the event will only increase her desire to go.

Having a pretext for getting together again and a plan set in stone also drastically reduce the chances that she'll flake. Even if she's not sure about you yet, she's more likely to come along anyway, just for the experience. Tagging along with a small group of interesting people to experience the best sushi in the world or check out the funniest comedian who ever lived or go to the coolest dive bar in town is a lot more tempting than just "going for coffee" or "getting together to talk sometime," which is how many guys ask women out. And compared to an actual date, in which she's trapped all night with a stranger with high expectations, your low-pressure event is a much more appealing option.

Make sure you avoid seeding with events that are complex, far away, or longer than a few hours. People are less likely to say yes to something if the cost of commitment is high.

Once you start seeding compelling plans into a conversation, the phone number exchange and the next meeting will occur effortlessly. Especially after you complete tomorrow's missions.

DAY

MISSION 1: The Way to Digits

The sole focus of today is the second part of the phone number exchange. So read your Day 20 Briefing and learn this useful, mostly wordless, and nearly rejection-free companion piece to seeding before moving on to Mission 2.

MISSION 2: Approach, Seed, and Exchange

Approach women today using the material you've learned so far.

Seed each conversation with a plan from your calendar, as you did yesterday.

If you hit the hook point, attempt the number close you learned today before ending the conversation.

Your mission is complete after you've either received one telephone number or approached five women. Whichever happens first.

DAY 20 BRIEFING
EXCHANGING NUMBERS

There are four things every Challenger should carry in his pockets at all times:

- gum or mints to eliminate bad breath.
- a pen to write down information.
- paper—ideally business cards, even if they're someone else's.
- condoms, because if you want to stay in the game, you have to play safe.

A lot of people collect digits on their cell phones, and that's okay. There are some fun routines for cell phone number exchanges, such as inputting a humorous phrase in her phone instead of your name, so that when you call, her display reads "Hot Tamale." But good old-fashioned pen and paper has many advantages, chief among them the following technique:

Yesterday you learned how to seed a plan into a conversation. The next step is to return to the topic when you're ending the interaction.

For example, just when the conversation is at a high point and you're about to leave, throw in something like the following, almost as an afterthought: "And make sure you check out the Sushi Nazi sometime." Short pause. "Actually, you should come along with us on Thursday because then I can finish telling you about the personality types we were discussing."

Note that adding an additional incentive to go—a "because" pretext, such as closing an open loop—further lessens the possibility of flaking or rejection.

Afterward, tell her, "Here, I'll give you my information." Women may have an autopilot response when guys ask for their number, but they'll rarely, if ever, object to taking your information.

Here's what you do next: Pull your pen and a business card (or some other small piece of paper, like a receipt) out of your pocket. Tear it in half. Then write down your name and number on one half.

Afterward, hold on to the scrap of paper with your number and hand her the blank half of the card along with the pen. She'll accept them; it would be rude not to.

Four times out of five, she'll write down her name and number. The few times when she doesn't, she'll ask, "What am I supposed to do with this?" Simply show her your half of the card with your information on it, and look at her with an expression that translates as "Duh, what else are you supposed to do with it?"

Now you have your information on your paper scrap and she has her information on her half. So just exchange the scraps. Fair is fair.

Visualize this movement and practice it a few times until it's natural and smooth.

It seems simple, and it's supposed to be.

The number exchange is not a magic trick. It won't make someone who has no interest in you suddenly give you her contact information. It's a tool to help you sail smoothly through an often awkward and precarious social ritual. I've never been rejected doing this, and I've never been given fake information. The reason is not necessarily the technique itself but the timing.

The key to making this work is simply to do it after you've hit the hook point. Once you've captured her imagination with your great conversation, flair, and personality, she'll be disappointed if you leave all of a sudden without exchanging contact information. So as long as you seem sociable and trustworthy, show her that you're more interesting or attractive than her other options, and don't try to exchange numbers too early, this transaction will proceed smoothly.

If you want to be a smart-ass—and I recommend it—once she's written down her number, tell her: "Draw a picture of yourself in case I forget what you look like." You'll be able to tell a lot about her from what she draws. Plus, it's fun.

Once you have the phone number, don't leave. Keep talking to her for a couple of minutes. If you just dash off, she'll think you were only interested in her for the number and she'll get buyer's remorse. Instead, after you've exchanged numbers, share one more anecdote to make her comfortable. If you don't know what to say, tease her about the self-portrait she just drew for you. "What's that supposed to be? An arm? Yeah, I think I see the resemblance."

Finally, remember that a phone number is not an end point in the game of attraction. It's just a resting place. In some cases, you may not need to get a phone number right away, because she'll want to spend the night with you. In other cases, you may get the phone number in the first fifteen minutes but spend hours together afterward. And every now and then, you'll make a definite plan to meet later that day and not even exchange phone numbers. Though men tend to treat obtaining a phone number like it's some sort of great victory, ultimately it's just a bookmark allowing you to pick up an interaction where you left off.

DAY

MISSION 1: Meet Your Silent Wingman

Today is an easy day.

It's also an important one.

Because today you will synthesize the information you've received so far and fit it into a larger framework of attraction, seduction, and courtship.

Your Day 21 Briefing includes a list of each step of the game you've learned, from opening a conversation to obtaining a phone number. Fill in the blanks with all the material you've successfully learned and used. When you're finished, add in any material you'd like to try. Then tear it out, photocopy it, or rewrite it on a regular sheet of paper.

Consider it your cheat sheet and silent wingman.

MISSION 2: Approach Using Your Silent Wingman

Take your completed cheat sheet, fold it, and put it in your back pocket.

Your goal today is to approach a woman (or a group containing a woman) and make it all the way from the top of the sheet to the bottom.

As long as you eventually get to the number exchange, it isn't necessary to use material from every category on your cheat sheet—or even most of them. It's simply your safety net.

As you master the game, you'll find that planned or scripted material becomes necessary only as backup, in case an interaction loses momentum or isn't progressing naturally toward the next necessary stage in creating a relationship. The best way to reach mastery is to add everything you can to your repertoire—and then, once you start experiencing success regularly, to remove as much as you can without affecting your results. In other words, practice using your cheat sheet, so that one day you'll no longer have to rely on it at all.

MISSION 3: General Courtship Strategy

What's the master plan? Perhaps it's time I let you in on it.

If you don't know where you're going, you won't know how best to get there. So turn to the second section of your Day 21 Briefing and read the article about the big picture.

DAY 21 BRIEFING
SILENT WINGMAN WORK SHEET

Attitude and Affirmations

I am relaxed, confident, playful, non-needy, unflappable, radiating positive energy. I will let go of my outcome. I am a man who women desire and want to be around. I will learn something from everyone I meet. I am testing women to see if they meet my standards. I deserve the best.

____

____

____

Openers

____

____

____

____

Roots

____

____

Time Constraints

Waypoint

"How do you all know each other?"

Disqualifications

Demonstrations of Value

Cold Readings

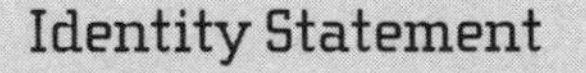

Identity Statement

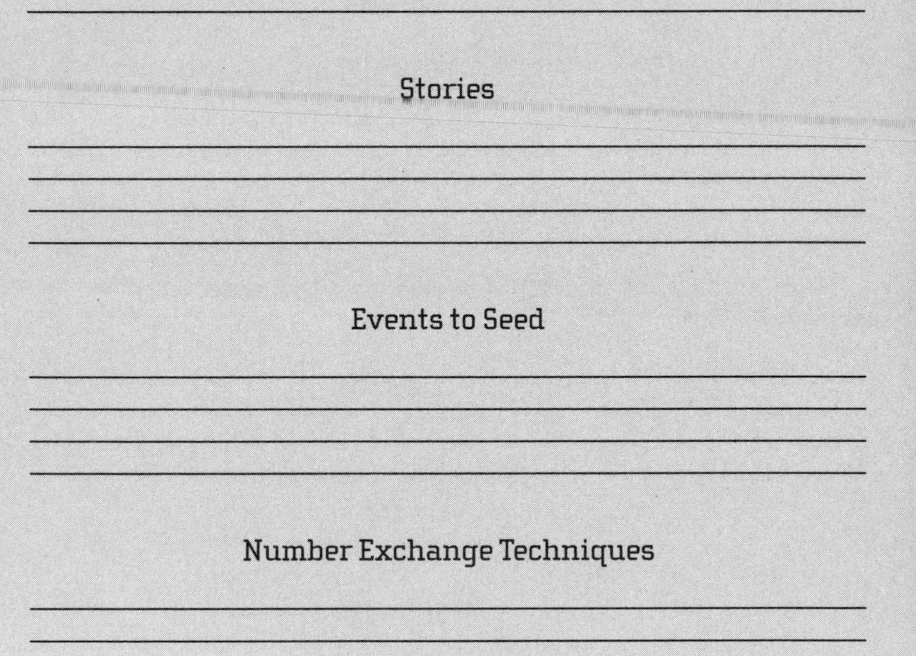

Stories

Events to Seed

Number Exchange Techniques

THE ANATOMY OF ATTRACTION

In the old days, my courtship strategy was simply to hang in there and be the last man standing. So I would make sure that either she was talking or I was talking at all times, and then hope that after enough hours and alcohol had gone by, I'd be able to make my move.

Once I worked up the courage to lunge for the kiss, though, I'd get the dreaded cheek turn. This was usually followed by a short speech explaining that she didn't want to ruin our friendship. It felt like a dagger plunging into my heart every time.

I couldn't figure out what I was doing wrong. I just thought I wasn't attractive or confident enough. And I'd repeat the same ineffectual strategy every time I had the opportunity to go out with a new woman, hoping that this one would like me.

When I discovered that attraction was a learnable skill, I quickly realized what should have been obvious to me the whole time: that every love story needs a plot. Two strangers must go through a specific sequence of events if a sexual or romantic relationship is going to build between them. And whether this sequence occurs through conscious effort or just naturally on its own, almost all relationships follow it.

I grew up thinking that one stage—building rapport—was the whole picture, which explains why I kept getting stuck in the friend zone. Friendships are built on rapport, trust, and common interests. What I didn't realize is that attraction can be built just as easily, but using different materials.

Once I understood this, everything changed. Eventually, as my interactions with women changed from friendships to romances, I was able to create a map and a clear route from the beginning of the courtship to the end. And as long as I knew where she was on that map and how to bring her to the next checkpoint, I no longer had to fear the dreaded cheek turn.

There were only five checkpoints:

1. *Open:* Every romance begins with two strangers meeting. This is how your parents met. This is how their parents met. And this is why the first nine days of the Challenge were dedicated to the minutiae of the approach, enabling you to break the ice in the most rejection-free way possible.

2. *Demonstrate value:* Once you've opened, your goal is to hit the hook point as soon as possible. Depending on the woman, her options, her self-esteem, and her interests and preferences, demonstrating value can involve as little effort as saying hello, or as much as making yourself seem like the most coveted person in the room while captivating her and her friends with powerful non-needy routines that display your worth and excellence.

3. *Create an emotional connection:* Sure, you're cool and interesting. But you could be talking to anyone in the room. Why her? It's time

to show that the two of you are bonded in some way, have things in common, click, understand each other, and were meant to meet.

4. *Structure a call to action:* Just because she likes you, that doesn't mean she's going to sleep with you. A window of possible intimacy has opened, but if you want her to jump through it, you'll have to give her an incentive to do so in the moment. Most commonly, this is done by arousing her through talk or touch. Time, comfort, trust, and laughter can also accomplish this. But sometimes she needs a stronger reason to make that physical leap. These techniques—eliciting jealousy, giving mixed messages, or even disappearing for a little while—will help her realize that if she doesn't move fast, she may lose her one opportunity to get together with you.

5. *Make a physical connection:* Once she's interested in going further, all you have to do is avoid making any mistakes that will cause her to change her mind—and walk with her across the bridge to physical intimacy in a way that doesn't make her uncomfortable, cause her to feel used, or elicit any other negative autopilot response.

Keep in mind that not every courtship starts at the beginning phases. Sometimes the interaction starts later in the process—if, for example, she's already attracted to you. In the future, you may even get to the level where you can sometimes walk up to a woman and make out with her within minutes. The better you get, the faster you'll be able to move through these stages.

A CLOSE-UP VIEW

The steps above helped guide me through nearly every approach I made. However, there are other ways to portray the same process. And different people respond better to different models.

So I sat down with the Stylelife coaches and asked them to come up with their own version for you, going into greater detail. There are six phases in their model. Here's what it looks like:

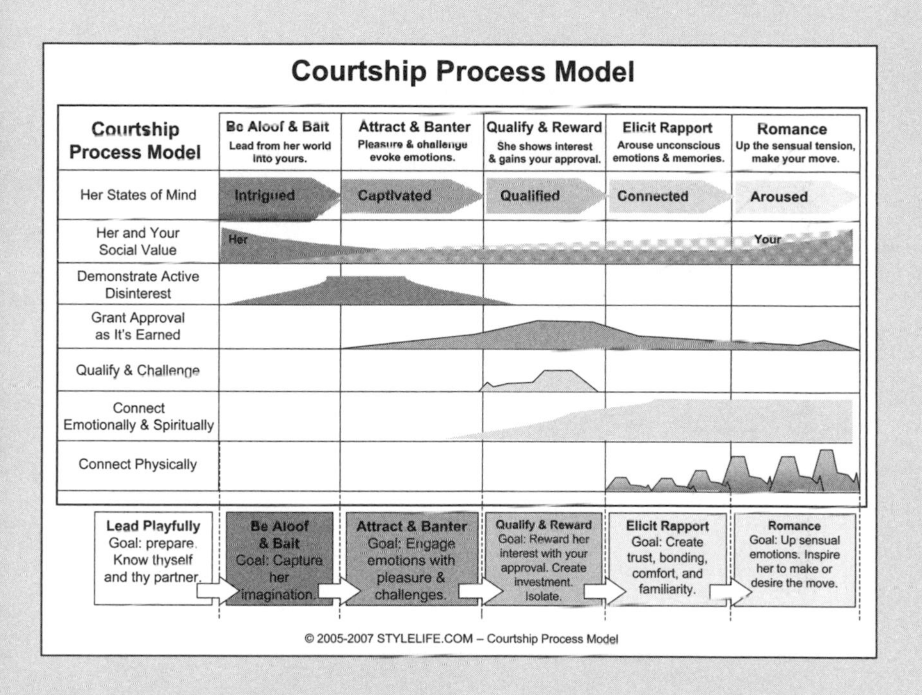

This model applies to both men meeting women and women meeting men. Each phase develops to an important milestone or turning point, allowing the relationship to advance to the next phase.

While understanding these phases in a developing relationship is helpful, knowing how to smoothly and successfully advance through them is much more useful. So I asked the team to break the phases into further detail and suggest specific actions to take and attitudes to have at each point in the process. Here's what they came up with:

Courtship Process Strategy

Courtship Phase	Goal State:	Strategy: What to do When
Self-image: Lead Playfully Goal: Prepare yourself. Know yourself, your ideal partner, and outline of the plan.	Confident	Develop your assets into your identity.
		Identify your demographic of potential partners.
		Develop and know yourself and your strategy.
		Master your inner self; be a playful leader.
Phase 1: Be Aloof & Bait Goal: Capture her imagination. Get her thinking about you as an integral part of her world.	Intrigued	Appear as the object of desire, but be aloof.
		Approach nonthreateningly, time constraint.
		Ready to leave, actively disinterested, disqualify.
		Create intrigue & curiosity, inspire her to engage.
Phase 2: Attract & Banter Goal: Create pleasure & challenges to engage her emotions. Generate attraction.	Captivated	Prove your social value, create intense emotions.
		Create light confusion, challenge, tease, banter.
		Reach the hook point, win over her friends.
		Demonstrate social proof, increase social value.
		Create opportunity for quality alone time together.
Phase 3: Qualify & Reward Goal: Reward her with your approval for her interest. Create investment in you.	Qualified	See her potential, challenge her efforts.
		Qualify and challenge.
		Reward, establish commonalities, show interest.
		Cold read, control frame, reframe if needed.
Phase 4: Elicit Rapport Goal: Create trust, deep bonds, comfort, and the feeling you've known each other in the past and future. Your encounter is special and meant to be.	Connected	Entertain with stories and games.
		Change locations, create new experiences.
		Demonstrate trust, relationship telescopes time.
		Elicit core values, recall rapport memories.
		Associate with positive feelings.
		Deepen bond and connection.
		Test kinesthetically and escalate.
		Insinuate and motivate so she pursues you.
Phase 5: Romance Goal: Turn up the sensual tension and physical emotions. Inspire her to make or desire the first move.	Aroused	Create a sensual atmosphere.
		Elicit sensual values & create erotic state of mind.
		Employ erotic kinesthetic teasers and escalate.
		Observe and respond to what turns her on.
		Make the bold move, indirect sensory explosion.
		Cuddle and chill together without rushing.
		No regrets; she feels good about her decision.
		Set and manage expectations.

You don't need to memorize all these phases and strategies, as long as you understand their subtext—that attraction isn't random, seduction isn't something that just happens, and courtship doesn't have to involve fumbling. The fact is, whether other men are using it consciously or not, there is a formula that makes a select few of them successful with women and in life.

You now have that formula.

DAY

MISSION 1: Learn to Flip the Script

Today is frame-control day, in which you'll learn techniques to stay dominant in a conversation. These concepts will not only be of use in nearly every social situation, but they may just change the way you look at the world.

Your first task: Read all about them in your Day 22 Briefing before proceeding to the rest of today's missions.

MISSION 2: Constructive Reframing

Your first mission is to reframe negativity into positivity at least once over the course of the day.

When you hear a friend, colleague, or stranger complain or say something negative, try to reframe it into something positive. For example, if a friend says that he's incompetent at something, tell him that he just likes to do things perfectly.

If someone says, "My girlfriend is driving me crazy," respond, "Why do you think she nags? It's only because she cares. If she didn't care, she wouldn't nag."

Keep reframing until the person accepts one of your positive conclusions.

If you don't hear anything negative all day, then call a friend or relative, ask what his or her biggest complaint or annoyance has been this week, and reframe that into something positive.

MISSION 3: Flirtatious Reframing

Choose from one of the following two flirtatious reframing exercises. Your mission is complete when you've performed it successfully one time. When you say these, make sure you're smiling and it's clear that you're not serious:

1. Reframe an accident into an intention: Go to a crowded place, such as a popular bar or store. When someone bumps into you or brushes against you as she walks past, jokingly say with mock indignation, "Did you just grope me? You know, I'm not that easy. I require dinner and a movie first."
2. Reframe kindness into self-interest: Go to a CD store and talk to a female employee or customer. Ask for advice on a good CD to play in the background at a dinner party—something new and cool. When she suggests a CD, teasingly accuse her of being paid to say it. "You really think I should get *that* CD? Hey, you're not getting a kickback from the record label, are you? You probably get, like, a washing machine or something for every hundred copies you sell." Then consider buying the CD. You'll find out why on Day 24.

MISSION 4: When the Going Gets Tough

If you haven't successfully exchanged phone numbers yet, study your silent wingman, put it in your back pocket, make sure your calendar is up to date, and approach four more women or groups today.

DAY 22 BRIEFING
CHANGING THE FRAME

By Thomas Scott McKenzie

An artist frames a painting. A carpenter frames a house. Project managers establish a time frame for getting work done. A criminal evades capture by framing a stranger. A film director frames a shot. Bowlers get ten frames a game.

There are dozens of different interpretations of the word *frame*, but most of them have to do with a structure or an agenda. In *Introducing NLP*, their classic book on neurolinguistic programming, authors Joseph O'Connor and John Seymour define frames as "the way we put things into different contexts to give them different meanings; what we make important at the moment."

In other words, a frame is the context through which a person, thing, or environment is perceived, and framing is a way that you can shape an interaction to achieve the result you desire. You can change your own frame, someone else's frame, or the frame in which a certain conversation or situation seems to exist.

Reframing is the process of changing the frame or providing a new view. "Reframing literally means to put a new or different frame around some image or experience," Robert Dilts writes in his book on the subject, *Sleight of Mouth*. "Psychologically, to 'reframe' something means to transform its meaning by putting it into a different framework or context than it has previously been perceived."

In fact, most kinds of flirting really amount to reframing. For example, if a woman bumps into you and you ask, "Did you just grab my ass?"—you've just reframed the situation from an accidental collision to a sexually charged situation.

Most social rules can also be thought of in terms of frames. The alpha male, for example, is the person with the dominant frame (or point of view) in a given situation. Dominance, however, should not be confused with being stubborn or a control freak. As Dilts asserts, "The person with the most flexibility will be the one who directs the interaction."

When you first meet a woman, it's important to have a strong frame, so that she feels a need to seek your approval, rather than the other way around. This is one of the reasons you're filling your Stylelife calendar with events: so that the woman can enter your world.

Even most of the things you're not supposed to do when approaching—such as acts of supplication, like buying a woman drinks so she'll talk to you—can be seen as evidence of having a weak frame or giving in to someone else's frame.

Reframing Techniques

Though there are innumerable techniques for reframing, in *Sleight of Mouth* Robert Dilts focuses on four specific ones.

CHANGING FRAME SIZE

Dilts uses the movie *Cabaret* as an example of how frame size affects our perception. One scene in the film begins with a close-up of "an angelic-looking young boy who is singing in a beautiful voice," he writes. But as the camera pulls back, viewers notice that he's dressed as a soldier. As it pulls back a little farther, viewers see his arm—and on it, an armband with a swastika.

"As the frame size gets larger and larger, we eventually see that the boy is singing at a huge Nazi rally," Dilts concludes. "The meaning and feeling conveyed by the image is completely changed by the information coming from the changes in the frame size of the image."

So during your interactions with women, imagine that you have a movie camera and can control the frame size. Let's say that you want a woman to leave the bar and go home with you, but she's worried about what her friends will think. Her frame is the equivalent of a group shot in your movie. You can zoom way out and tell her that her time on this planet is short, that adventures she'll always remember are awaiting her, and that if she constantly inhibits herself based on the opinions of others, life will pass her by. Or you can zoom into a close-up, cutting her friends out of the picture and focusing on just her wishes and desires, creating an intimate world between the two of you that she doesn't want to leave.

CONTEXT REFRAMING

Context reframing is based on the fact that the same event will have different implications depending on the circumstances or environment in which it occurs. "Rain, for example, will be perceived as an extremely positive event to a group of people who have been suffering from a severe drought, but as a negative event for a group of people who are in the midst of a flood, or who have planned an outdoor wedding," Dilts writes. "The rain itself is neither 'good' nor 'bad.' The judgment related to it has to do with the consequence it produces within a particular context."

This is useful to your inner game as well as your outer game. Let's say that you've just tried a new opener, but the woman gave you a funny look and walked away. In the context of trying to get a phone number, you would view the interaction as a failure. But if you reframe the context so your goal wasn't to obtain the digits but to determine the effectiveness of your new opener, then the interaction was a success.

CONTENT REFRAMING

Content reframing acknowledges that people see the same thing differently based on their personal attitudes, likes, dislikes, needs, and values. Dilts uses the example of an empty field of grass. A farmer sees it as an opportunity to plant crops, an architect sees it as a lot to build a Gothic home, a man flying a small plane that's running out of fuel sees the field as an emergency landing strip.

We all see things differently. Reframing based on content means looking at each individual's perspective and the intention behind his or her external behavior.

So, suppose you're back at that bar with the woman you want to take home. But her friend keeps telling her, "You guys should just stay here. Why do you need to go anywhere else? You shouldn't leave with a guy you just met."

It would be easy to simply dismiss the friend's behavior as selfish and controlling. But try to find a positive intention in her actions. Maybe she's worried about her friend's safety. Maybe she thinks you're the kind of guy who drives a van with garbage bags taped over the windows and power tools banging around in the back.

She may seem hell-bent on frustrating you, but her behavior is actually coming from a positive place. And the quicker you understand her frame, the better you can handle the objection. For example, you can deal with the situation by spending some time talking with the friend so that she trusts you more, and then giving her your phone number. This way, if she's worried about her friend or wants to find out where she is, she has the option of calling you.

REFRAMING CRITICS AND CRITICISM

The problem with critics is that they don't just point out what you're doing wrong. They often point out what they think is wrong with you.

To deal with critics, it's important to get beyond the negativity and realize that their judgments are usually made with good intentions.

This also applies to your criticisms of others. When a friend offers an idea, for example, avoid responding with something negative that could start an argument like, "That'll never work." Instead, ask a positive, constructive question that he or she won't take personally, such as "How are you going to pull that off?"

This type of reframing also works well on your fiercest critic: you. Take any excuse you may have that keeps you from achieving your goals, such as "I don't have time," and turn it into a solvable problem: "I don't use my time efficiently." Then turn that problem into a question: "How can I use my time more efficiently so that I can reach my goal?"

Reframing criticisms and limitations as "how" questions can turn a dead end into an open door.

Framing The Game

The more you learn about frames, the more flexibility, fun, and success you'll have in your social and professional life. At the very least, always keep in mind the following three things when interacting with women:

1. Always keep a strong frame. Have her meet you in your reality, rather than changing yourself to fit into hers. More than money and looks, this attitude will help you convey status.

2. Reframing is the key to both persuasion and flirtation. It gives you control of a conversation, with the ability to redirect it somewhere humorous, positive, exciting, or, at the right time, sexual. Practice it as much as you can, and not only will you become more successful with women, you'll become a more talented speaker and better-rounded thinker as well.

3. Use these techniques in moderation. Do not become obsessed with controlling the frame in every interaction all the time. Sometimes surrender can be victory.

DAY

MISSION 1: Self-assessment

Welcome to your final review day.

Below are a few of the skills you've learned so far. Rate yourself by circling a number from 1 to 10 in each area, with 1 being completely deficient, 5 being average, and 10 being perfect in the skill or trait listed.

Posture	1	2	3	4	5	6	7	8	9	10
Vocal Projection	1	2	3	4	5	6	7	8	9	10
Vocal Tonality	1	2	3	4	5	6	7	8	9	10
No Vocal Pausers	1	2	3	4	5	6	7	8	9	10
Grooming	1	2	3	4	5	6	7	8	9	10
Clothing Style	1	2	3	4	5	6	7	8	9	10
Inner Game	1	2	3	4	5	6	7	8	9	10
Eye Contact	1	2	3	4	5	6	7	8	9	10
Energy Level/Positivity	1	2	3	4	5	6	7	8	9	10
Approaching Strangers	1	2	3	4	5	6	7	8	9	10
Using Openers	1	2	3	4	5	6	7	8	9	10
Time Constraints	1	2	3	4	5	6	7	8	9	10
Rooting	1	2	3	4	5	6	7	8	9	10
Disqualifiers	1	2	3	4	5	6	7	8	9	10
Expressing a Unique Identity	1	2	3	4	5	6	7	8	9	10
Demonstrating Value	1	2	3	4	5	6	7	8	9	10
Non-neediness	1	2	3	4	5	6	7	8	9	10
Storytelling	1	2	3	4	5	6	7	8	9	10
Cold Reading	1	2	3	4	5	6	7	8	9	10
Spontaneous Conversation	1	2	3	4	5	6	7	8	9	10

Open Loops	1	2	3	4	5	6	7	8	9	10
Seeding	1	2	3	4	5	6	7	8	9	10
Exchanging Numbers	1	2	3	4	5	6	7	8	9	10
Frame Control/Dominance	1	2	3	4	5	6	7	8	9	10
Reframing	1	2	3	4	5	6	7	8	9	10

Select the areas in which you ranked yourself the lowest and work on those today, using the material and exercises already provided.

The final dash to get a date begins next week, so make sure you're caught up.

MISSION 2: Get a Lifeline

If you still haven't received a phone number, that's okay. One of two things is probably happening.

The first is that you've hit a sticking point. If so, it's time to get a helping hand. Go to www.stylelife.com/challenge and enter the Challenger forum. Start a thread there with the title "Sticking Point." Discuss the specific area where you're having trouble, providing as much detail as possible. Using the advice you get from coaches and fellow Challengers online, make four more approaches today.

The second possibility is that you've just been reading the book and haven't been doing the field assignments. Shame on you.

If you have already received a phone number or been on a date, don't just sit there and gloat. Go out and make four more approaches as well. Practice makes perfect.

MISSION 3: Start Persuading

Now that you know what works when meeting women, it's important to understand why these techniques work, so that you can best respond to the fluctuations, surprises, and unexpected circumstances that occur in nearly every social situation. So turn to your Day 23 Briefing, read the book report on *Influence* by Robert Cialdini, and fill in the blanks.

DAY 23 BRIEFING

THE ENGINE OF YES—A BOOK REPORT

In *Influence: The Psychology of Persuasion*, psychology professor Robert B. Cialdini examines the shortcuts that people use to make decisions, then distills the tactics of persuasion to six key psychological principles.

Cialdini's focus is on sales and advertising. However, his principles help explain not just what makes people buy a particular car or brand of soap, but also how people make decisions about each other.

Below is a brief summary of Cialdini's principles. Each has scores of applications to the process of creating attraction. For example, the principle of social proof explains why women are more attracted to men who are accompanied by other women than men who are alone. After each principle, write down at least one practical way you could employ it to improve your game.

A word of warning: These are powerful principles, and they should be used to appeal to the nobler side of people, not to their weaknesses. Steer people in the direction of their own best interests, not just yours.

Social Proof

This is the principle of majority rule: If a lot of people are doing something, others tend to believe it must be the right thing to do. As Cialdini explains, "One means we use to determine what is correct is to find out what other people think is correct."

Social proof is particularly persuasive, he notes, when the person trying to make a decision is uncertain or in an unclear situation. It's also more powerful when the individuals we're observing are people we relate to or believe are just like ourselves.

APPLICATION: ____

____

____

____

Liking

Perhaps the most obvious of them all, the principle of liking holds that we're more inclined to agree to the requests of someone we know and like.

Cialdini cites several factors that produce liking. These occur when someone has a similar fashion style, background, or interest as us, gives us compliments; is physically attractive; or has repeated contact with us, especially in situations where we have to cooperate with him or her to achieve a mutual benefit.

Cialdini adds an interesting twist to this principle: "an innocent association with either bad things or good things will influence how people feel about us." For better or worse, he continues, "If we can surround ourselves with success that we are connected with in even a superficial way . . . our public prestige will rise."

APPLICATION: ______________________________

____

____

____

Reciprocation

If people do something for us, we feel obliged to pay them back. Even "people who we might ordinarily dislike . . . can greatly increase the chance that we will do what they wish merely by providing us with a small favor prior to their requests," Cialdini writes.

An interesting corollary, he adds, is that in order to get someone to agree to a small request, a good tactic is to start by making a large request that he or she is likely to turn down.

APPLICATION: ______________________________

____

____

____

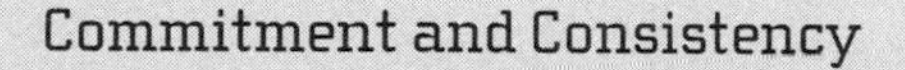

Commitment and Consistency

When people make up their mind about something, they tend not to change it—especially if they back it up with an action or a statement. Even when confronted with facts to the contrary, they often won't change their decision or belief.

"Once we have made a choice or taken a stand," Cialdini explains, "we will encounter personal and interpersonal pressures to behave consistently with that commitment."

There are many corollaries to this rule. One is that people often observe their actions in order to determine their beliefs, instead of letting their beliefs guide their actions. Another states that if you can get people to commit to the decision to buy something, but the price rises or the rules change before they have a chance to purchase it, they'll still want it. And, finally, there's the foot-in-the-door technique: To get people to commit to a large purchase, have them first make a small, inconsequential one.

APPLICATION: ____

____

____

____

Authority

This principle states simply that we tend to be obedient to authority figures, even at times when their wishes make no sense or conflict with our personal beliefs.

One side effect of this, Cialdini notes, is that we're as suggestible to people who merely possess symbols of authority as we are to legitimate authorities. The symbols we often kowtow to include professional titles; uniforms or formal attire; expensive status symbols; and commanding or convincing speaking voices. We even tend to accept as an authority someone who's simply larger than us.

APPLICATION: ____

____

____

____

Scarcity

According to the rule of scarcity, people perceive things that are rare, or becoming rare, as more valuable and desirable than they would if they were readily accessible. "Opportunities seem more valuable to us when their availability is limited," Cialdini notes.

One of the most important conclusions Cialdini draws from this is that "the idea of potential loss plays a large role in human decision making." Thus, when obstacles are placed in the way of something or our access to it becomes limited, our desire for it becomes greater. We then tend to assign more positive qualities to it in order to justify the desire.

"Because we know that things that are difficult to possess are typically better than those that are easy to possess," he writes, "we can often use an item's availability to help us quickly and correctly decide on its quality."

He adds that we tend to desire objects whose availability is suddenly restricted, more than items that have always been scarce.

APPLICATION: ____

____

____

____

The Next Level

The most powerful motivators occur when different principles of persuasion join forces—for example, when social proof combines with scarcity. "Not only do we want the same item when it is made scarce," Cialdini writes, "we want it most when we are in competition for it."

For your final exercise, write down one example of how two different principles can be combined to create a strong motivator for attraction.

APPLICATION: ____

____

____

____

DAY

MISSION 1: Be the Party

One of the biggest mistakes men make when trying to make plans with a woman is not having a plan in the first place. "I don't know. What do you want to do?" just may be the worst way to ask a person out.

The next worst way is asking her, "So what are you doing on Saturday?" And then inviting yourself along.

Rather than trying to glom on to her lifestyle, a better frame to have is that perhaps she's not getting everything she wants from her life and is hoping to step into someone else's exciting world. And that world just happens to be yours.

The Stylelife Challenge is not just about women, it's about lifestyle. If you can build a positive, exciting orbit of people, places, and things around yourself, one that other people respect and want to be a part of, you will meet and attract women automatically.

So to close out the Stylelife Challenge, you are going to plan a dinner party for Day 30. Your task is to read today's briefing and find out just how to pull this off before moving on to Mission 2.

MISSION 2: Seed Your Stylelife Party

Your mission today is to seed your dinner party.

Approach women and groups using the material you've learned. But instead of seeding an event in your calendar, seed your dinner party. You may want to discuss the theme or occasion for the party, and mention any friends who share something in common with her. But don't invite her.

Only when the conversation is ending, and it's time to exchange numbers, will you invite her to the party.

One way to do this is to say, "You know what? You should come to the dinner party. I think you'll really enjoy some of the people there. And, besides, we need a wild card."

If she asks what a wild card is, either tease her by saying "someone unpredictable" or compliment her by saying "someone new and interesting." What you choose to say here depends entirely on her self-esteem.

Unless she's really excited about going, don't give her the details of the party on the spot. That can come across as too eager. Wait to talk on the phone first. This way she'll have to work a little harder for it, and demonstrate that she's trustworthy and will mix well with your friends.

Your mission is complete after you've either collected the phone number of one potential party guest or made five approaches. Whichever happens first.

Tomorrow you will be using that number.

DAY 24 BRIEFING

YOUR STYLELIFE DINNER PARTY

Do you know what's great about having a party?

It's an excuse to get the phone number of nearly anyone you meet, as well as an excuse to call anyone you haven't talked to in a long time. No number will ever go stale as long as you have the occasional dinner party.

For the purposes of the Challenge, the definition of a party is simply six or more people gathering in any public or private location for the purposes of a fun, recreational, bonding experience.

Intent

Having a dinner party allows you to get together with a woman on your turf, where she has to compete for your attention. It also makes for an easy, low-commitment date. There are plenty of people around to keep the conversation going and build the anticipation you both will have for private time together later.

Furthermore, having a regular party will add to your circle of friends and potential girlfriends; build your social skills; strengthen your leadership qualities; and help you develop the kind of lifestyle others want to be a part of. Some of the most desired women in the world don't just date actors, musicians, directors, billionaires, and athletes, they also date club owners and promoters. This is because everyone wants to be accepted by the in-crowd. So do them all a favor by creating an in-crowd and accepting them.

Promotion

You don't need to create invitations for your dinner party. And, whatever you do, don't make flyers for your party. This is a small, exclusive event with a hand-picked guest list, and flyers imply mass, indiscriminate invitations.

You do, however, need a reason for having the party. It doesn't need to be anything complicated. Consider presenting your party to women as a weekly ritual where you gather some of the most interesting people you've met for good food and conversation. Or, better still, actually make it a weekly or monthly ritual. You could call it Monday martini night or the Tuesday charades challenge or the Wednesday international cook-off. If you want to get really pretentious, you could even call it a salon.

Another option is to create an occasion for the dinner. If a friend of yours has done anything of note—released an independent CD, published an article, started a website, had a birthday, adopted a puppy, bought a new shirt—throw a party to honor him or her. Then play the new CD, read an excerpt of the article, or proudly display the new shirt at the party.

Another pretext is to make it a holiday. Every single day on the calendar commemorates something—national sibling day, barbershop quartet day, the birth of Gary Coleman—so throw a party to celebrate.

Location

Your party can take place at any of a number of locations.

The best venue is your house or apartment, or the house or apartment of a friend. There are only a few necessary preparations you need to make: cleaning the space, providing something to eat, selecting appropriate music, and—assuming you and your guests are of legal age—having enough alcohol to last throughout the party.

If cooking isn't your forte, a dinner party can be an excuse for you to learn. If one of the women you've met enjoys cooking, convince her to help out. Since your guests know you're throwing the party to teach yourself to cook, they won't even mind when the turkey catches fire. As long as there's alcohol to drink.

If you don't have the time or incentive to cook, just order out food, remove it from the to-go containers, leave it warming in the oven until the guests arrive, then serve it in regular dishes. If no one asks, you don't need to tell them it's from the Greek restaurant down the street.

If the gathering is fewer than ten people, provide an enclosed sitting space to facilitate conversation. Buy cheap folding chairs if you have to. If you're less experienced in hosting, start or end the night with a group event, such as a favorite weekly television show or an interactive game like charades. Never underestimate the appeal of anything that was fun at age seven.

The second-best venue is a lounge or restaurant that has tables or couches large enough for your entire group. Make a reservation in advance and confirm it on the day of the party. It's perfectly fine for everyone to split the bill. Though in reality it's no different from a regular dinner out, your intent to celebrate as a group is enough to justify calling it a party.

Other locations include a park or beach for an evening picnic or barbecue; a bar or club; even a bowling alley, hotel room, or amusement park. Your only limits are your imagination and the law.

Casting

You're not going to throw some kind of blow-out keg party, unless you really want to. Most likely, it'll be a small dinner party for a select group—and that's how you're going to explain it to the woman you're talking to. The more select and exclusive your party appears to be, the better it will turn out and the quicker word will get around town.

For example, rather than saying you're inviting people, tell her that you're "casting" the dinner party—picking and choosing just the right combination of interesting personalities, interests, and occupations—and she might make a good addition to the cast. After all, every party needs a wild card.

Though calling her a wild card can be a fun tease, you actually will want one at your party. So make sure you invite someone whose conversation or behavior is slightly eccentric and outgoing (but not unpleasant or extreme). It takes the pressure off you as a host, because the guests will have someone else to talk about and entertain them.

You'll also want to invite at least one male friend who's a good conversationalist, at least one female friend or couple, and the women you've met (or will meet) during the Challenge. It is crucial to make sure there's more than one woman present at your party, so that the girl you're interested in doesn't feel uncomfortable or outnumbered.

If more than one woman you met during the Challenge shows up, don't worry if they compare notes on how they met you. Just keep your frame strong: You're a social person who enjoys going out and meeting new people, discussing whatever's on your mind with them, and bringing them together to network. If you live in this reality, they'll usually end up competing for you.

If she wants to bring a friend, don't panic. Let her. If you charm her friend, you're likely to charm her as well. Even if it's a male friend, that's okay. After all, you've invited other women, and those women can even help keep him occupied. Though you don't want to encourage her to bring friends, if she does, it will only widen your social circle and make the next party that much better.

If you're having the dinner party at a home, the energy can sometimes dip after the meal. One way to prevent this from happening is to invite a second shift of four to eight people for cocktails afterward. The new faces, enthusiasm, and energy will give the party the spark it needs to make it lively and memorable. (Be careful about the timing: most guests arrive roughly a half hour after the time you tell them the party starts.)

For each person you invite, make sure that you have an interesting way to introduce him or her—consider using the same kind of identity statement you made for yourself. The better you make your friends look, the better you look.

Connecting

There are several things you can do before, during, and after the party to deepen your connection with the woman you're interested in.

If the party is at your home, have her stay and help you clean up. If the party is elsewhere, think of somewhere to go afterward.

Sometimes it can be fun to get her involved in helping out with the party, rather than just relying on your work and hospitality. To do this, give her tasks or assignments, such as bringing or cooking food.

One friend of mine makes sangria with his dates. The work is light, it involves alcohol, and it's perfect for two people. To do this, get a bottle of Spanish wine, two limes, two lemons, two oranges, a mango, and a half cup of sugar. Pour the wine in a pitcher, let it breathe for ten minutes, then add the sugar. Squeeze the juice of a lime, lemon, and orange into the wine. Have her slice up the rest of the fruit into wedges and add it to the drink. Refrigerate it for an hour if possible, toss in a tray's worth of ice cubes, and pour it for your guests. (This recipe serves five people, so double it if you have ten guests.)

Other activities to do together range from shopping for ingredients (the grocery store can be a fun first date) to attempting to roll your own sushi, which can get messy—and that's a good thing.

Be careful not to dote on her too much or bend over backward to keep her entertained. And don't get jealous if another guy at the party starts talking to your date. As the host, you're the man of the moment; no one is a threat to you. If you have a trusted friend, let him know your identity statement, so he can share his admiration of you with your date.

The goal of the dinner party is to have a good time, build an exciting lifestyle, and bring together people who will find one another interesting. If you can accomplish this, the attraction will take care of itself.

MISSION 1: Phone Rules

The step after exchanging numbers—calling—is a source of anxiety for some men. However, the rule of phone engagement is simple: Don't do anything wrong. She's only just met you, and one warning signal is the only excuse she needs to decide never to see you again.

You don't want that to happen. So your task is to read the Day 25 Briefing on phone game.

MISSION 2: Plan Your Party

If you haven't settled on a location for your dinner party on Day 30, do so.

Write down your ideal guest list of six to ten people below. Include any women you've exchanged numbers with. Write each person's name in the column on the left and his or her identity in the column on the right. Your description of each person should be terse and compelling, so that when you scan this list, the party looks like a special event.

	Name	Identity
1.	______________________	______________________
2.	______________________	______________________
3.	______________________	______________________
4.	______________________	______________________
5.	______________________	______________________
6.	______________________	______________________
7.	______________________	______________________
8.	______________________	______________________

MISSION 3: Harvest Your Seeds

Phone all the women whose numbers you've collected in the last few weeks. Practice the telephone techniques you learned in your briefing.

Invite each woman to the event or party you've planned for Day 30. Make sure you give her a specific location and time to arrive. Emphasize that it's going to be a small, handpicked group, so she knows her invitation is a privilege and her presence is crucial to the mix.

Compared to asking complete strangers for movie recommendations on the phone, this should be a piece of cake.

If you haven't yet received a phone number, make five more approaches today, with the goal of party recruitment in mind. Make sure you study your cheat sheet first.

If you've already been on a date, don't forget to add your name to the winner's circle on the Stylelife forum and share the story with your fellow Challengers.

DAY 25 BRIEFING

PHONE GAME

You know, I used to wait two days to call anybody, but now it's like everyone in town waits two days. So I think three days is kind of money. What do you think?

—FROM THE FILM *SWINGERS*

So you've had a successful approach and exchanged numbers with a woman you really like, but now what? What if she's forgotten you? What if you're too nervous on the phone and blow it? What if she's busy on the day you want to see her? What if she's in the middle of something more important when she answers? What if a guy picks up the phone? What if she's given you a wrong number? What if California falls into the ocean?

Don't worry about it.

If you relax, the first phone call can be a very simple process.

How Long to Wait

How long are you supposed to wait between getting the number and making the call?

Some say phone the next day; others say wait three days.

They're all wrong. There is no fixed amount of time that needs to pass.

Rather, here's how long you can wait: as long as you possibly can.

In other words, if you meet a woman and make an amazing connection, and she begs you to call her, you can wait as long as a week. She's not going to forget you.

However, if you meet a woman, talk for a few minutes, exchange numbers, and afterward see her talking to different guys all night, you're going to have to call her the next day. This is because, if you didn't make that deep a connection or impression, within forty-eight hours she's likely to have forgotten all about you.

When it comes to call times, the general rule is: Don't lose the momentum. Call her while the interaction is still fresh in her head, but not so soon and so often that she thinks you're a stalker.

To Block or Not to Block?

Many so-called experts advise blocking your phone number when calling a woman. They also suggest that you not leave a message if she doesn't pick up.

The idea, they explain, is that if you keep calling, eventually she'll answer—and once you've trapped her, you can convince her to see you.

I don't use or recommend this crowbar method, unless you're a telemarketer.

The fact is: If she's not calling you back or taking your calls, the problem is not your phone game; it's your approach game, because you didn't convey the qualities necessary for her to want to see you again. In fact, whenever something goes wrong at one stage in the interaction, it generally means you made a mistake in the previous stage.

So never block your calls and always leave a message. Why? Because it shows confidence. If you displayed an attractive personality, demonstrated your value, and conveyed trust when you first met her, she's going to be excited when you call.

Your goal should be to leave every interaction with the woman worrying, "What if he doesn't call?"

If you've seeded your event properly, when you do phone, she'll know just what you're calling about and she'll be comfortable taking the call.

What to Say

Here's a general structure to follow on the first phone call:

1. Try to avoid introducing yourself by name. Instead, begin the conversation with a callback to your previous conversation. If you used the Village People opener to meet her, when she picks up, say slowly and confidently, "So I found out: There's no fireman in the Village People." She'll know who it is. If you teased her by calling her a brat, when she picks up, just say, "Hey, brat." This way, instead of reminding her that you're a stranger (especially if she happens to have forgotten your name), you bring her back to the good time she originally had talking with you.

2. To avoid an awkward pause, after she greets you, launch into a quick story from your life. Select an appropriate narrative you created on storytelling day, or add a new one to your repertoire. Begin by saying something like "The most amazing thing happened to me today . . ." Just make sure your story is short, and that the point of it isn't to build yourself up but to make her smile, laugh, and feel comfortable.

3. Speak in a deep, calm, comfortable voice tinged with fun and positive energy. It's good to be upbeat, but don't talk too fast or be too hyper. Smile on the phone, and she'll hear it.

4. After telling your short story, give her a chance to speak. Most of the time she'll tell you about her day or ask a question. If she doesn't, just move on.

5. Make plans for later in the week. Some experts suggest saying what days you're busy first to demonstrate, among other things, that you have a full life and are squeezing her into it. Incorporating the push-pull you learned on disqualification day, you might say something like "I'm busy Friday and Saturday, but I'm having a small dinner party on Sunday. I'm casting a group of really interesting people, and you should come. We need a troublemaker."

6. If you're inviting her to an event other than your party, don't frame the interaction as a date. Invite her to "hang out," "tag along," or "join" you and your friends.

7. If she says she can make it, great. If she's busy, let her know about one of the other events on your calendar. And only one. Unless she gushes with enthusiasm to go, tell her that she'd probably enjoy it and if a space frees up, you'll call her and let her know.

8. Whether or not she's available, don't suddenly say good-bye and hang up after inviting her out. Just as you did after exchanging phone numbers, continue the conversation for another minute or two. Add a little playful banter or share a quick, related story.

9. End on a high note. Be the person who says good-bye first. You're busy. You've got things to do.

Though this script is simple and has been used effectively by thousands of men, it's not the only way to handle the first phone call. As you become more comfortable with the process, you may want to distinguish yourself from other men by calling first just to talk briefly and then making plans on the second call.

If you prefer to text, try to avoid it for your first interaction. On the other hand, if you fall into the trap of phone tag before having your first conversation, texting can save the day.

If She's Too Busy Again . . .

If she's vague about committing to plans or turns down multiple invitations, it's time to examine your game. At some point in the initial interaction, you probably made a mistake. Perhaps you conveyed lower social value, came across as desperate, or exchanged phone numbers too early. Maybe your sense of style (or lack thereof) didn't fit her dating criteria. Figure out what your shortcoming was and work to improve it. In a few rare cases, if you're doing everything right but she's still flaky, she may have a boyfriend or be getting over one.

In general, never accept the words "too busy" as a regular excuse. If Angelina Jolie called and invited you to a dinner she's having at her mansion with Bono, Jay-Z, Bill Clinton, and George Lucas, would you be able to make it?

Of course you would. You'd break whatever plans you had, blow off work, and probably walk there on your hands if you had to.

Your goal in every interaction is to be so interesting and such a rare find that she's never too busy for you. After all, if you met the perfect 10, wouldn't you manage to find time for her?

So be the perfect 10.

DAY

MISSION 1: Clear Your Mind

This may be the most challenging day so far, but it will also provide the greatest benefit to your intuitive understanding of the game.

Your first task: Forget everything you've learned so far.

MISSION 2: Approach Unarmed

Approach three women or groups today—using no material.

Do not start the conversation by asking for an opinion. Do not use scripted disqualifiers. Do not discuss rings and Greek gods. Do not tear business cards in half.

Improvise something—perhaps about someone around you or an item she's wearing or whatever's on your mind at the time—to start the conversation. Don't be afraid of small talk; asking generic questions about work, movies, and travel; or even buying her a drink if you're in a bar or a café. Break all the rules.

Stay in the conversation until she excuses herself, or it's clear that she wants you to leave. It may get awkward, but hang in there.

If possible, time the interaction. Your goal is to stay in the conversation for at least ten minutes without using material.

If all goes well, feel free to invite her to your dinner party or one of your calendar events.

MISSION 3: Live the Difference

Reflect on your approaches today.

Did you notice any differences between using material and freestyling? Any differences between how you interacted before the Stylelife Challenge versus now? If so, write them in the space below.

____

____

____

____

____

MISSION 4: Fill-ins

Your final task today is to read the following advice on filling in the gaps:

There's a sticking point that some Challengers hit around this point. They approach a group, open, demonstrate value, cold read—they do it all. Yet inside they feel tense and awkward, because they have no idea what to do *between* all these techniques. What do they say? How do they transition from piece to piece? How do they get to a point where they can exchange phone numbers?

These are, of course, irrational fears—after all, they've managed to have interesting conversations with people before. Overcoming material dependency, and realizing that you have plenty of things to say to fill in the gaps is one of the goals of today's field assignment.

It can be easy to forget that it's your personality, more than the material, that will make her want to see you again. Routines are great because they show you to be more interesting than most guys. They also serve as springboards to get you to the next stage in an interaction. But your entire conversation doesn't need to be one big performance. You don't want the woman to think of you as a monkey in a little hat, turning the crank on a music box for her entertainment.

So stay up-to-date on entertainment, culture, current events, and happenings around town; cultivate the ability to pay attention to the details of what other people do, say, and wear; master the art of social intelligence; get comfortable in your own skin; and, if you're still having a problem filling in the gaps, take improv comedy courses to learn spontaneity.

If the game is self-improvement, then we're all in it for life. So learn to play it right.

DAY

MISSION 1: Learn to Connect

Imagine if you met a woman whose favorite musician and film were exactly the same as yours; who shared your strongest beliefs and opinions; and who turned out to have grown up just a block away from you, even though you'd never met. Wouldn't you feel like you'd met someone incredible?

This is the power of rapport. And it's something you want to create with every woman you're interested in. So turn to your Day 27 Briefing and read about it before moving on to the rest of today's missions.

MISSION 2: Date Your Calendar

Print or copy a fresh Stylelife calendar page.

Fill in events—as well as selling points and reasons to go—every day until the end of the Challenge. Make sure you include your party. Then familiarize yourself with the activities, the dates you listed them on, and the reasons for going.

MISSION 3: Rapport Workout

Choose two of the three rapport exercises below to perform. It's okay to try them with a coworker, cashier, casual acquaintance, or even in an online chat, but you'll get more out of the exercises if you do them with a new person or group you've approached. If the interaction goes well, make sure you invite the woman you're interested in to your dinner party or one of your calendar events.

Pay close attention to the other person's reactions as you increase and decrease your level of rapport.

MAKING AND BREAKING RAPPORT

During the following exercise, observe the other person's reaction as you instantly create rapport—and then quickly break it.

Have a conversation like the following:

YOU: Where are you from?
HER: [*Whatever city*].
YOU: Oh my God, no way! I grew up there too. What school did you go to?
HER: [*Whatever school*].
YOU: Get out of here. I went there too.
HER: Really?
YOU: No. I've never actually been there. [*Then, in a dry monotone*] Are you upset?

RAPPORT TEST

In the following exercise, break rapport and then see if the person you're talking to will strive to re-create it.

YOU: Out of curiosity, what's the last CD you bought or song you downloaded?
HER: [*Some song by some artist*].
YOU: Really? I'm surprised. I'm not the biggest fan of their music.

If she backpedals and says she doesn't really like the artist either, this means she's seeking rapport with you. If she tells you why she likes the music or disagrees with you, then either she isn't seeking rapport or she's simply confident in her taste and opinions.

PHYSICAL RAPPORT

This exercise illustrates the power of body language to affect someone else's state.

During a conversation with someone you're comfortable with, cross your arms and turn away from them while they're talking. If seated, cross your legs away from them as well. Remain in that position for a minute or two.

See if the person starts to get rattled or uncomfortable—or even comments on it. Then uncross your arms, open up your body language, and turn toward him or her again. If the person is a good friend, ask if he or she noticed or felt a difference when you broke physical rapport.

Repeat this exercise one more time today with a different person.

DAY 27 BRIEFING
THE PATH TO RAPPORT

Creating rapport is the process of developing a connection with someone based on trust, comfort, commonalities, and affinity. For many men, it's the easiest and most natural part of the courtship process.

Rapport is the point in the interaction when she sees those little parts of you that you try to hide sometimes—your inner nerd, your goofy side, your enthusiasm for superhero comics, or musical theater, or monster truck rallies—and finds them endearing. It's the moment when she shares her innermost thoughts, experiences, and feelings—and you understand them, perhaps better than anyone else she's ever met. It's when you find yourselves laughing in unison or starting to say the same thing at the same time.

In short, rapport is when two people really get to know each other and find out that, yes, they were supposed to meet. How lucky they must be.

At the same time, rapport is a castle built of Lego. It can be dismantled in an instant and put back together a few seconds later. Knowing when and how to build and break rapport will help propel an interaction through the stages necessary to create a romantic or sexual relationship.

Watch any love story. Before two lovers fully unite, they first lose rapport—maybe through a misunderstanding or a disapproving parent or a spurned rival or punishment for a mistake. They experience anguish, and then, in their sorrow, realize just how strongly they feel about the other person. Only when rapport has been regained and mutual feelings confessed do they feel complete again.

So-called nice guys make the mistake of trying only for rapport with a woman, to the exclusion of everything else that builds attraction. There's a fine line between naturally having rapport and being seen as trying too hard to get rapport.

In addition, timing is key. If you strive for rapport too early, the relationship may fall into the friend zone. If you strive for rapport too late, she may think you're a player who doesn't see her as the dynamic individual she is. The best time to build rapport when meeting a woman is after hitting the hook point but before getting too physical. Now that she's interested in you and invested in the interaction, you can even ask all the questions you were advised not to when first meeting her.

To help you create the kind of rapport that magically just happens, I've asked Stylelife senior coach Don Diego Garcia to break it down.

And he did, into two neat categories: lead and sync.

Lead

For decades, parents trusted their children to be entertained by Fred Rogers through the TV program *Mister Rogers' Neighborhood.* He started each show with a mild manner and a friendly "Hi, neighbor!"

Notice that he didn't say, "Hi, stranger!" He *assumed* you were his neighbor. Although you probably never lived anywhere near Fred, he made you feel as though you did. Mr. Rogers assumed a neighborly affinity and went about his show as though you were an old friend in his living room. It was a hit.

Though you don't want to be as exaggeratedly friendly as Mr. Rogers, you do want to assume rapport with women in a similar way. To do so, simply ask yourself this question: "How would I act if this person were a lifelong friend?" Now pass that answer through a filter of social propriety, and you'll know how to approach.

You should assume rapport from the moment she first sees or hears you. Suppose there's someone you want to meet in the dairy section of the local supermarket. An approach that assumes formality begins with you holding out your hand and introducing yourself by name.

An approach that assumes rapport, however, begins differently: "I could understand 2 percent for people who can't decide between whole and nonfat milk, but 1 percent? Is there really that big a difference between 2 percent and 1 percent?"

People also bond naturally with credible leaders who possess such qualities as confidence, authority, authenticity, security, self-assuredness, courtesy, and honesty. Staying grounded in these qualities will prevent you from succumbing to the risks of seeking rapport—such as supplication, losing your frame, falling into the friend zone, or becoming her therapist instead of her lover.

Sync

Carl Jung liked to talk about synchronicity as attaching meaning to events that are coincidental. I call the process of actively producing this state syncing.

Syncing is not copying or imitating everything your partner does. Syncing is a more subtle form of falling into pace with them and cultivating empathy. People in groups do it unconsciously all the time. When you sync correctly, your partner will bond with you more on an emotional, spiritual, and energetic level than on an intellectual level.

Let's examine the ways you can get in sync with the woman you're interested in.

VISUALLY

To sync visually with a woman, watch her posture, facial expression, breathing pace, gestures, or even blinking rate, and match them. Remain relaxed and calm as you do this. If you match her just right, she'll start subconsciously mirroring your body language as well.

AUDITORALLY

If you notice that she uses a few specific words frequently or that certain words seem to have a special meaning to her, consider them hot-button words and mentally store them for future use. You can also match your language to her work jargon, regional expressions, and any words that define her as a member of a particular subculture.

Auditory syncing can also involve paying attention to words that suggest that the speaker has a special affinity for certain senses. For example, visual people tend to use words like *focus, bright, see,* and *show* when discussing their thoughts and desires. People who live in their feelings use words like *touch, feel, aware,* and *sense*. Audiophiles prefer descriptors like *ring, sounds,* and *click*. Listen closely to her speech patterns, pick up on which sense words she uses, and then sprinkle them into your own conversation.

You can also match other things about her way of speaking—her pitch, volume, tempo, timbre, or tonality—or even her nonverbal utterances, from groans to laughter to pauses. This may sound extreme, but it's practically common sense that a slow talker and a fast talker, for example, will have a hard time communicating. The slow talker will have trouble following the fast talker, and the fast talker will be impatient with the slow talker. The more similarly you communicate, the more likely you'll get along.

LOGICALLY

Sync logically by discovering particular interests, aesthetics, morals, sensibilities, or background details you have in common. This common form of building rapport involves playing the "me too" game. Me-too topics can include family experiences, travel stories, career goals, entertainment preferences, personal idiosyncrasies, and relationship criteria.

You can sync logically with light rapport topics: where she's from, why she's out, what her interests are. Later in the interaction, move into deep rapport, using morality conundrums, personality tests, imagination exercises, confessions of vulnerability, intimate stories, and discussions of goals and dreams.

In a nutshell, similarity leads to affinity. Affinity leads to rapport.

EMOTIONALLY

As you're talking to the woman you're interested in, wholeheartedly invest yourself in understanding how she thinks and feels. Master the skill of empathy to put yourself in her place. See things from her point of view. We all want to find someone in this big, alienating, often uncaring world who understands us.

US VERSUS THEM

One of the most powerful ways to build rapport is to create a conspiracy in which only you two have something in common, and no one else gets it. This can range from bonding over a peculiar idea that few others believe to role playing and telling others that you're childhood friends or even engaged. These latter gambits are particularly powerful because the roles themselves are ones of increased rapport.

Troubleshooting

Though some of these subtle strategies may take a conscious effort at first, eventually they'll become more automatic. The best way to master them is to practice one at a time until you understand how each works. You'll notice, for example, that mirroring her breathing will subtly change the energy around the two of you and draw you closer together to the exclusion of everyone else in the room.

Often, the biggest barrier to creating wide and deep rapport is not the other person but you. If you're too scared to reveal yourself or show any vulnerability, then she typically won't feel comfortable letting down her guard with you. Rapport is a two-way street. And it doesn't exist without trust and openness.

So if you ever find it difficult to achieve rapport, whether it's because of your masks and walls or hers, consider letting your guard down, forgetting about all these techniques, and just relating to her with an open heart. You may be surprised.

DAY

MISSION 1: Your Internal Compass

There is one key piece of the game that most people never mention, teach, or realize exists. Even if you stop using routines and abandon the structure you've been taught, you will still be relying on this.

Beyond its uses in attraction, this is a skill that affects all areas of your life, whether you're interviewing for a job or getting held up at gunpoint.

Read about it in your Day 28 Briefing before proceeding with the rest of today's assignments.

MISSION 2: Are You a Psychic or a Psycho?

The following exercise works best with a seated group of two or more people who look easygoing.

Your assignment is to guess how they know one another. Are they related? Roommates? Friends from work or school? In a relationship? On a date? Taking a class together?

Make an educated guess. Then walk up, ask, and find out if you're correct.

Your calibration skills will not only help you guess correctly, they'll also help you pose the question in a way that doesn't make the group feel like it's part of a laboratory experiment.

For example, you can say: "You have to help me quickly settle a debate I was just having with my friend. We noticed you all talking, and he said you guys probably all work together. I guessed you were friends from college."

If they give you a funny look—which will happen occasionally—acknowledge the oddness of the situation by saying something like, "I know, strange question, but he's into psychology. He does this stuff all the time. Then I have to do the dirty work."

Make sure that you're smiling, your approach comes from a place of healthy curiosity, they know you're not asking in a judgmental way, and you use a time constraint.

Your mission is complete once you've approached three groups or made one correct guess, whichever comes first.

If the conversation goes so well that you end up joining the group for a while, take the opportunity to stock your dinner party with some new faces.

MISSION 3: Get Proof of Interest (Optional)

If Mission 2 seems too easy, or if you want to do more calibration training today, then here's an additional goal to add to your approaches above.

Your secondary mission is to receive at least one indicator of interest from a woman in one of the groups you approach. Study the list of attraction signals in today's briefing to familiarize yourself with these indicators.

If you don't receive an indicator of interest in one of the three groups you approach today, then make two more approaches using your standard opener.

Your mission is complete once you've received one indicator of interest, or you've approached five women or groups altogether today.

If you do receive any indicators of interest from a woman you've approached, then it's your duty to exchange numbers and invite her to your party.

DAY 28 BRIEFING

CALIBRATION

There are only three things you need to perfect in order to master the art of attracting women:

- Who you are
- What you do
- When and how you do it

When it comes to who you are, during the first few days of the Challenge you worked on your goals, mission statement, and identity. Tomorrow you'll drill down and refine the individual characteristics of your personality.

As for what you do, you've spent nearly every day developing that element of your game, from openers to demonstrations of value.

And, for when and how you do it, you've learned the order and sequence of each attraction event and studied the big picture. But there is one more piece to this puzzle: calibration. And it makes all the difference.

Technically speaking, calibration is the act of adjusting or correcting the accuracy of a measuring instrument, usually through determining its deviation from a standard. In terms of attraction, the definition remains the same—but the measuring instrument is you and the standard is her.

Identifying the Instrument

When approaching, calibration is the skill that allows you to read the dynamics of the group or the woman you're interested in and know what to do next.

If, for example, a woman saunters up to you in a bar, rubs your chest, and says you're cute, what do you do?

If you try an opinion opener, you'll bore her—and a demonstration of value will seem like you're trying too hard. Through calibration, you'll know to skip most of the stages you've learned and start thinking about how to give her the physical experience she's looking for. Further calibration will help you determine if she wants to make out with you right there, if she wants to be taken home, or if she's just trying to make someone else in the room jealous. All these evaluations—made in a fraction of a second—will determine your next course of action.

Calibration continues to be necessary throughout an interaction. Making slight adjustments in your body language, eye contact, and tonality can affect the behavior, responses, and interest level of the woman you're talking to. Try standing too close to her and noticing how she reacts; then stand too far away. Try leaning in, then leaning back. Explore making direct eye contact, looking at her mouth, or looking over her shoulder when talking.

Learning to read her responses, and then adjusting your actions to elicit the feelings you want her to have, is the core of the game.

Setting the Instrument

Though calibration is one of the most critical pieces of the game, it can also be a trap. If you overcalibrate and worry too much about every small sign a woman gives you, you'll probably become anxious and insecure, and sabotage the interaction.

When meeting a new person, all kinds of thoughts and snap judgments, both positive and negative, may swim through your mind in a matter of moments. So to avoid erring on the side of insecurity, when you're trying to assess how she feels about you, set your calibrator not to 0 (neutral interest) but to +2 (slightly interested). Go into every interaction with the attitude that the woman you're interested in is into you—and if you find yourself wondering how to interpret something she does, assume the best. This will motivate you to press forward with confidence.

Labeling the Instrument

After setting your instrument this way, you should then try to determine how she currently feels about you and what she needs to progress to the next stage in your attraction sequence.

At all times, you're looking for one of three responses from her:

- Green Light—A positive response, which means go forward
- Yellow Light—A neutral response, which means proceed with caution
- Red Light—A negative response, which means stop what you're doing

Red lights are the realm of damage control, when you've miscalibrated and crossed a line or made an error. If that occurs, back up to the last yellow light.

The yellow light is what you'll encounter most often. It's a point when anything can happen. And the outcome depends on your ability to assess where she is in the courtship process, where she needs to be taken next, and what she needs to get there. Among the things she may need from you are more value, more attraction, more comfort, more trust, or just more time.

Make these calculations in your mind as imperceptibly as possible. One bad habit people sometimes develop while learning the game is that they become reaction seeking. Remember, as soon as it becomes clear that you've done or said anything solely to get a particular response from her, it not only loses its impact, it also appears needy.

The game hinges on subtleties and details like these, in part because, whether she knows it or not, she's also calibrating you. And most women have far more finely tuned instruments and intuition than we do.

Reading the Instrument

Some people's calibration is a little off. They can't seem to tell when they're making people uncomfortable—or, conversely, when a woman is actually attracted to them.

No matter where you currently stand, if you pay attention and learn from the feedback a woman gives you, you'll accumulate enough experience and success that your calibration will correct itself. Eventually your intuition will become so strong that you won't need to apply any rules to calibrate. You'll just know.

In the meantime, here are a few clear signs that can help you tell whether a woman is attracted to you. These signals are subtle, so don't rely on just one to give you the green light to proceed. Make sure you have three to four clear, positive indications before assuming she's interested in getting a little more intimate. These indicators of interest include:

- She asks you, without prompting, what your name is, what you do for work, where you're from, or how old you are.
- You lean back, and she leans toward you.
- Her legs are uncrossed (or crossed toward you), her body is angled toward you, and her arms are uncrossed.
- She changes her opinion of a song, movie, or current event based on your opinion.
- You make a joke, and no one in the group laughs but her.
- You take her hand to lead her somewhere, and she squeezes it—especially if you let go and she holds on.

- She says, "I'm not going to sleep with you" or "I'm not going home with you," before you've asked her to or conveyed any intent to do so.
- She playfully punches or slaps your hand or arm.
- She ignores her friends when they try to contribute or want to leave.
- You stop talking and make eye contact, and she holds it for longer than a second.
- You turn to speak to someone else, and she waits for you to turn back to her.
- She displays a combination of subconscious attraction gestures: lip licking, hair twirling, pupil dilating, even nostril flaring.
- She grooms herself or adjusts her clothes to expose more skin while talking to you.
- She absentmindedly fondles something like a straw, cell phone, or piece of jewelry. (If she's clutching it tightly or fidgeting with it, that's not a good sign.)
- You stop talking, and she tries to continue the conversation, usually with the word "So . . ."
- She mirrors your movements—stroking her hair after you stroke yours, sipping her drink after you sip yours, even making a face after you make one at her.

Like sending out a sonar signal and waiting for it to return to determine a distance, you can send out signals to test her interest. To do this, make a small action and notice how she responds. For example, playfully (and lightly) punch her in the shoulder. If she punches or hits you back, these are good signs; if she stiffens or recoils slightly, these are not so good.

Be forewarned that some women will be very touchy-feely as soon as they meet you because they seek the validation of guys chasing them, enjoy the power it gives them over men, or are showing off for someone else in the room. With these women, don't consider anything a genuine display of interest unless you know you've earned or deserved it. Until then, tell them with a smile that you charge twenty dollars a touch, and they're racking up quite a bill.

Upgrading the Instrument

We've discussed calibrating to determine your course of action. But there's another type of calibration that's more fun and powerful. It includes elements of cold reading, determines which disqualifiers are appropriate if any, and helps build rapport.

Like using X-ray glasses, this advanced form of calibration allows you to explore her innermost thoughts, needs, and desires. To train yourself to do this, as you're talking to her, ask yourself:

- What type of personality does she have?
- Does she have high or low self-esteem?
- Is she sexually open or reserved?
- What does she do for work?
- Is she currently in a relationship?
- Is she an oldest, youngest, middle, or only child?
- Is she closer to her mother or her father?
- Is she primarily athletic, emotional, or intellectual?
- What qualities is she attracted to in men?
- What are her needs?
- Where is she in life and what is she looking for?

Just like with cold reading, there are many clues that will give you this information. They include her clothing, makeup, posture, gestures, eye movements, the way she speaks, and the people she's with.

Mastering the Instrument

There's only one way to master calibration: Get feedback.

The simplest way to practice is to turn on a soap opera and watch it with the sound off. Try to guess as much as you can about the relationship between the characters on-screen. Then turn on the volume and check your accuracy.

A good intermediate exercise is to make polite, informed guesses about new people you're talking to. Try to determine what they do for a living, what kind of environment they were raised in, whether they were popular in school, and what their birth order is. Then, at some point during the conversation, ask and see if you were correct.

Once you're comfortable doing this, next time you're out with friends, look at a group of two or more people and figure out as much about them as you can. In addition to the details already discussed, try to determine their relationship to one another, if they're local or visiting, and what their general story is.

When you're finished, simply walk over and ask them if you're right. Make sure you smile, ask with genuine curiosity, don't make them uncomfortable, and don't seem like you're making fun of them or judging them. Not only will this give you the feedback you need to improve your calibration, rapport, and cold-reading skills, but it's a great opener—as you'll discover in your field exercise today.

DAY

MISSION 1: Step on the Scale

As you learned yesterday, there are three aspects to the game: who you are, what you do, and when and how you do it.

Today we're going to further explore the idea of who you are. It's not easy to make lasting improvements to the characteristics of your personality, but once you begin the process, you'll start moving toward your goals in dating and life as if you were on autopilot. You won't need to whip out the rings routine to demonstrate value, because you'll be demonstrating value simply by existing.

The switches of attraction and desire can be flipped by eight major personality attributes working together. Turn to your Day 29 Briefing, read about them, and rate yourself from 1 to 10 in each category.

If you have been doing the Challenge with a friend, told anyone about your missions, or found a local wing in the Stylelife forums, when you're finished scoring yourself, ask your trusted acquaintance to give you an honest rating in each category as well.

MISSION 2: The Final Sprint

If you haven't yet been on or arranged a date this month, it's time to make it happen.

If you haven't yet received a definite, ironclad confirmation for your dinner party from at least one of the women you've met, it's approach time for you as well.

Tomorrow the Stylelife Challenge ends.

And you have the tools it takes to be a winner. All you have to do is use and implement them.

To make sure no one gets left behind, I've saved one technique for today: the instant conversation starter.

Grab a notebook or a piece of paper. Write at the top, in capital letters, "TOP TEN FILMS." Now number it from one to ten.

Your mission today is to make a list of the top ten films of all time. You're going to play one or two of these in the background, with the sound off, at every party you have. Of course, with such an important task, you're going to need some assistance.

So go to one of the following five locations, where you're most likely to meet friendly, open-minded women:

1. A health-food grocery store such as Trader Joe's or Whole Foods Market
2. The lobby, lounge, bar, or pool area of a major hotel
3. The bookstore, library, cafeteria, or student center of a college
4. A spirituality bookstore, alternative coffeehouse, or yoga studio
5. An event from local newspaper listings that attractive single women are likely to attend, whether it be a wine tasting or a casting call

Make sure that you bring your list and a pen. Fill in five spaces anywhere on the list with film titles. But make sure you leave the number one and number two spaces blank for her valuable input.

Here's a sample script you may want to use: "Hey, you look like you may know something about movies. I'm trying to figure out the top ten films of all time for this weekly movie party I'm starting, and I'm experiencing total brain freeze. Here's what I have so far."

Then show her the list and have her help you fill it out. To disqualify, tease her for choosing frivolous or obvious movies; to create rapport, bond on favorite films. When the energy begins to flag, start a new thread by using an opinion opener, the rings routine, a story from your files, or anything else you've learned this month.

Your goal, of course, is to seed your party, invite her to it, and exchange phone numbers. Since this is the penultimate day of the Challenge, spend as long as it takes until you have a solid number exchange.

This is the first day of the rest of your dating life.

DAY 29 BRIEFING

WHO YOU ARE: THE L.A.S. V.E.G.A.S. SYSTEM

Rate yourself in each of the qualities below on a scale of 1 to 10, where 1 is completely deficient in the trait, 5 is average, and 10 is perfect. Judge yourself not as you see yourself but as you believe others see you. Try to be as honest and realistic as possible. Write your answers in the spaces below.

Looks

At the beginning of the Challenge, you learned that looks have less to do with your physical features than with how you present yourself. Rate yourself on your grooming, posture, eye contact, whether you stand out in a positive way, and if your style attracts the type of women you want to be with.
RATING: ______________

SUGGESTION FOR IMPROVEMENT: Study and execute more Day 5 tasks; find role models whose style you admire; make dates to shop for clothing, shoes, and grooming supplies with women you meet.

Adaptability

Ever notice that uptight men tend not to do well with women? This is because they aren't adaptable. Rate yourself on your adventurousness, spontaneity, independence, risk taking, social intelligence, flexibility, and ability to handle new situations and environments.
RATING: ______________

SUGGESTION FOR IMPROVEMENT: Write down a few things you'd like to do in your lifetime. Focus not on career or relationship goals but on recreational skills and adventures—learning to scuba dive, taking a safari, building a kit car, or competing in a triathlon. Then circle one of these items and commit to doing it in the next six months. Enter it into your calendar six months from now to make a firm deadline for yourself.

Strength

Strength is the ability to protect a woman and make her feel safe. Some men display this through money or muscle, but those aren't necessary—and often aren't enough. So rate yourself on being an effective communicator, having a powerful frame, living in your own reality, your ability to take care of others, and criteria such as assertiveness, leadership ability, courage, loyalty, decisiveness, and self-assurance.
RATING: ____________

SUGGESTION FOR IMPROVEMENT: From the list above, select one attribute you need to work on in order to add a point to your strength. Then start demonstrating it in social situations, whether it's showing you're decisive by ordering for a table of friends at a restaurant or demonstrating your communication ability by talking your way into a store when it's about to close.

Value

As you learned on Day 14, value is one of the key criteria people look for when deciding whom to align with. Value actually consists of three elements: what you think your value is, what she thinks it is, and what impartial observers think it is. Rate yourself on the degree to which you're the leader of a social circle, admired by others, able to teach people things, and comfortable displaying high-status behaviors. Other criteria include being intelligent, interesting, talented, entertaining, successful, self-sufficient, and creative.
RATING: ____________

SUGGESTION FOR IMPROVEMENT: Make a list of five reasons a woman would want to see you again after meeting you for fifteen minutes. The list should be based on the value you either project or provide to her. Commit to learning one new skill, game, or attribute to add to that list.

Emotional Connection

This is the home of rapport and abstract concepts like chemistry. It's about possessing qualities that make people feel excited, connected, comfortable, and understood around you, as if they've just met a best friend or soulmate. Rate yourself on your success in finding commonalities with strangers, creating deep rapport with people, being in touch with your feelings, listening closely to others; and on criteria such as compassion, positivity, selflessness, and empathy.

RATING: ____________

SUGGESTION FOR IMPROVEMENT: Fear, insecurity, and lack of self-awareness block the ability to emotionally connect with others. Try to spend part of each day communicating, sensing, and existing with an open heart and through your deepest feelings—whatever that means to you. Drop any pretensions, masks, and walls that separate you from others. If you disagree with people, rather than trying to make your point, empathize with what they're feeling. If you're not the type to meditate, then step outside your comfort zone, go to a class or retreat, and try it.

Goals

As discussed on Day 1, goals are defined not by what you do but by your ambitions and what you're capable of doing. Rate yourself on the clarity of your goals, dreams, and hunger for life. You can measure your potential to achieve them by determining if you possess traits like stability, efficiency, perseverance, and the ability to learn quickly.

RATING: ____________

SUGGESTION FOR IMPROVEMENT: Review the goals you set for yourself on Day 2. On a separate sheet of paper, write an actual timeline for achieving each goal, with definite benchmarks. Make sure you include any financial requirements or potential complications in your calculations. Adjust this schedule every year based on new insights, information, and accomplishments—and live by it.

Authenticity

An authentic person is happy with himself and embraces even his imperfections. Rate yourself on your congruence—the alignment between the face you show to the world and what you're really like on the inside. Keep in mind that having contradictory sides to your personality doesn't make you incongruent. Having a duality, contradiction, or complications can make you more rich and compelling as a person. But being phony, insincere, or disingenuous does not.

RATING: ____________

SUGGESTION FOR IMPROVEMENT: On a piece of paper, write down the qualities you try to portray to the world. Next to each, add a 1 to 10 rating for how closely that quality matches who you really are deep inside. For any quality you rated under a 7, write down the obstacle that prevents it from being true. For example, if you want others to think of you as confident, but you rated your actual feelings of confidence as a 5, then your obstacle is insecurity. If the trait is financial success, then the obstacle is your lack of wealth. Work to remove that obstacle. Sources of help can include self-improvement books, seminars, therapy, or life changes such as a new job, hobby, or social circle. This will not be a short or easy road, but you won't regret taking it.

Self-Worth

This may be the single most important attribute here, and the wellspring from which most of the others flow. Rate yourself on your sense of confidence and worthiness, as well as your lack of fears and insecurities about yourself. Examine your willingness to take up space as you move through the world, how well you accept compliments, how comfortable you are when other people pay attention to you, and how much you deserve the devotion of a woman of the highest caliber. Do you truly believe that you're entitled to the best the world has to offer?

RATING: ____________

SUGGESTION FOR IMPROVEMENT: In the end (and you're only one day away), self-worth is what the Stylelife Challenge has been all about. Don't stop learning and improving after Day 30. Continue to examine yourself rigorously, work on your shortcomings, eliminate sticking points, raise the bar for yourself, and develop relationships with positive-minded people. As you experience more and more success, you will recognize, embrace, and exude more and more self-worth.

Total Score (all eight categories) ________
L.A.S. V.E.G.A.S. score (total points divided by 8) ________

In the months to come, your long-term mission is to boost your L.A.S. V.E.G.A.S. score. It's much less work to attract the best when you truly are the best.

DAY

MISSION 1: Party Time

You're too busy to handle a big mission today. After all, you have a dinner party to throw. Refer to your Day 24 Briefing if you need any help making the final preparations and arrangements.

If you weren't able to invite any women to the party—or you're not sure whether enough people are going to show up—set aside a few hours earlier in the day to make approaches.

Go to a nearby location such as a mall, café, or other area where women gather. Make as many approaches as possible. As soon as you hit the hook point with a woman or group you like, give yourself a time constraint and invite them to your dinner party. If you go home without having recruited any extra guests, don't cancel your party. It's a great opportunity to strengthen your social circle and leadership skills.

Once party time comes around, don't panic if the guests roll in late. It'll all work out great. Enjoy it. Make sure the woman you're interested in is comfortable, but don't pay too much attention to her at the expense of being a generous and enchanting host. Make sure everyone's glasses are always full.

After dinner, if all is going well with your date, ask her to stay behind and help clean up. If your party is at a restaurant or public location, have a second place in mind to go afterward—an interesting bar, lounge, or event on your calendar. If you both drove there, suggest taking one car. This way, you can have some alone time with her.

Consider having this kind of party every week or every month, so that you can begin building a lifestyle that consistently attracts the kind of women you deserve to be dating.

MISSION 2: Congratulate Yourself

Congratulations. You made it to the final day of the Stylelife Challenge.

If you've performed all the previous assignments and feel you've improved yourself in any way this month, then you are a winner. Some people go through their whole lives in darkness.

If you got a date, be proud of yourself for meeting the Challenge objective. If you'd like to share the experience or get feedback on it, describe the details of your approach and date in the Stylelife winner's circle: www.stylelife.com/challenge.

If you didn't get a date, despite completing every mission, then you get an additional assignment today. Go to www.stylelife.com/challenge and listen to the audio lesson titled "Works in Progress." You may find a solution there to whatever held you back.

MISSION 3: Commit to Greatness

So what are you going to do on Day 31 and all the days after that?

Look at how much you improved in a month. Now just imagine the results you could get if you committed to the game for another month, two months, three.

There's still a lot left to learn: what to do on the date; the fundamentals of attraction; techniques of arousal; crossing the physical divide; handling different environments; turning friends into lovers; being more fun; manufacturing chemistry; applying persuasion; leadership; group dynamics; isolation; kissing; winging; body language speed-reading; inner-circle sexual techniques; and hundreds of great routines and advanced concepts. Everything you've learned so far is only the beginning.

The art of social dynamics is much like working out: If you stop going to the gym, your muscles begin to dissipate and return to their former size. So your penultimate mission is to go to www.stylelife.com/Day31 to receive a game plan for the future.

This may be the end of the road on the Stylelife Challenge, but it's the beginning of a new journey.

I'll see you on that journey.

MISSION 4: Into the Looking Glass

Your final task: Look at yourself in the mirror.

Who do you see?

Even though I spent years undergoing an intensive campaign to improve myself, sometimes I'd look in the mirror and see the guy who was never popular and never had a date in high school looking back at me. Despite appearing and acting completely different, I still sometimes saw the world through his eyes.

Similarly, some Challengers I've met went through radical transformations. They looked cool, had good jobs, dated amazing women, and were fun to be around. But when they looked in the mirror, they saw the person they used to be.

So if you don't love, value, and appreciate the guy looking back at you in the mirror, then it's time to change your lenses. I'm not going to ask you to see your true self in the mirror; few of us have that kind of perspective. But instead of seeing the old you in the looking glass, try seeing the person you're becoming. You're going to like him a lot more.

Remember, perception is reality. And when you see yourself as a guy who's socially awkward, you'll act that way and others will treat you that way—no matter what your external appearance and value may be.

But when you see the fun, positive, confident, graceful, socially savvy person you're becoming in the mirror, and consequently start seeing the world through his eyes, people will respond a lot differently to you—because you've just fought the hardest battle and won. You beat your old programming.

So clean up and take a good look at yourself in the mirror. Reflect back to when you examined yourself in the mirror on Day 4 and think about everything you've learned and accomplished since then.

Be aware of your posture, smile, and energy as you look in the mirror. Recall your most successful approach and the way the woman genuinely enjoyed you. Once you see your best self confidently beaming back at you—the guy who any woman would love to be around—take a mental snapshot of that guy. And carry that photograph in your head wherever you go. Because that guy is you.

Welcome to your new reality.

ACKNOWLEDGMENTS

The Stylelife Challenge is the result of lessons from thousands of approaches, years of camaraderie with the master pickup artists from *The Game*, feedback from students around the world, hundreds of books and research papers, and the contributions of the Stylelife Academy coaching team.

There are two contributors, in particular, who deserve special recognition. You've met them already in your briefings:

Don Diego Garcia is a San Francisco–based Stylelife Senior Coach with a heart of gold. He has written scores of the most highly regarded missions and ebooks in recent memory, positively influenced the lives of thousands of students, and helped proofread this book.

Thomas Scott McKenzie is a Midwest-based Stylelife Senior Coach and ace author. He has written for many journals and magazines, from the profound (*Tin House*) to the profane (*Stuff*). In addition to contributing to the briefings, he also helped edit the original multimedia Stylelife Challenge materials into these narrow pages.

Thanks also to Dessi, Haze, Organizer, Masters, Julia Caulder, Maddash, DJ, and especially Phoenix and Rourke for helping out behind the scenes on the earliest incarnation of the Challenge. Stylelife Coaches Evolve, Tommy D, Gypsy, and Bravo also helped make this book possible. The Challenger known as Exception deserves credit for the Village People opener mentioned on Day 18. And Rourke and Michal Gregus also contributed material.

Special thanks also to the seduction gurus who have influenced my life and this book with their teachings, and camaraderie. They include Mystery, now a star of stage and screen; David DeAngelo, who has branched out into the business world; Ross Jeffries, the father of the movement that spawned this mad-

ness; Swinggcat, the wizard behind the curtain; and Juggler, a great writer and, now, a married man.

Then there are two men whose names I cannot mention. They are part of a future book. But I owe the idea for the Stylelife Challenge to them. You'll read about them then. But I'd feel remiss if I didn't give them their due. So thanks to . . . those two guys.

The proofreading team consisted of many of the aforementioned characters, along with Anna G., Ersin Pertan, M the G, Todd Strauss, Dr. M. J., Nicole Renee, Aimee Moss, Kelly Gurwitz, Lauren, Evelyn Ng, and Sarah Dowling. Soa Cho and Kristine Harlan did the fact-checking and research, unearthing psychological and scientific papers supporting everything from the time constraint to the L.A.S. V.E.G.A.S. attributes to approaching with a smile. Thanks also to Drew Huskey and Neel Vora for organizing the core street teams.

My most enthusiastic contrafibularities go to the world-class HarperCollins construction team: Carrie Kania, Michael Morrison, David Roth-Ey, Lisa Gallagher, Rachel Romano, Chase Bodine, Cassie Jones, Brittany Hamblin, Michael Signorelli, and Cal Morgan, the fastest editor in the East. Thanks also to Judith Regan, who originally suggested turning the Challenge into a book.

Finally, I'd like to thank you for completing the Challenge and taking control of your reality. The only thing better than hearing the success stories is seeing the before and after photos. You guys are putting *Body for Life* to shame. Respect.

THE ROUTINES COLLECTION

“All the world’s indeed a stage,
And we are merely players.”

—RUSH

BY WAY OF SHAKESPEARE

THIS VOLUME IS NOT FOR EVERYONE.

THE FOLLOWING PAGES HAVE BEEN ADDED TO THIS COMPENDIUM WITH EXTREME MISGIVINGS.

THEY CONTAIN ABOMINABLE WORDS. CHEESY WORDS. FALSE WORDS. WORDS FROWNED ON BY MEN AND WOMEN ALIKE. THEY HAVE BEEN MISINTERPRETED, MISUSED, AND CONDEMNED. THEY HAVE TURNED THE LIVES OF THOSE WHO RELY ON THEM INTO BAD THEATER.

YET EVERY ONE OF THEM HAS LED ME TO SCORES OF FRIENDSHIPS, RELATIONSHIPS, AND MUTUAL GOOD TIMES.

SO YOU DECIDE: IS IT POSSIBLE TO HAVE A REAL RELATIONSHIP IF IT STARTS WITH A MEMORIZED, SCRIPTED ROUTINE?

IN MY EXPERIENCE, YES. BUT I DARE NOT TELL ANYONE THAT. THEY WOULDN'T UNDERSTAND THAT THESE WORDS ARE JUST TOOLS THAT ENABLE THE BEARER TO LIFT ONE OF THE HEAVIEST OBSTACLES IN THE WORLD: PEOPLE'S RESISTANCE TO TRUSTING A STRANGER.

IF YOU HAVE THE NATURAL STRENGTH TO OVERCOME THAT RESISTANCE YOURSELF, THEN YOU DON'T NEED THESE. IF YOU ARE STILL LEARNING, AND YOU'VE ALREADY GONE THROUGH THE MATERIAL IN *THE STYLELIFE CHALLENGE*, THEN HERE ARE A FEW MORE EXERCISES TO HELP BUILD YOUR SOCIAL MUSCLES.

JUST REMEMBER THAT THESE ARE ONLY TRAINING WHEELS. BECAUSE THE SECRET OF THE GAME IS THAT THE BETTER YOU ARE AT IT, THE LESS YOU HAVE TO USE IT.

CONTENTS

INTRODUCTION

There is no such thing as a pickup line.

But there are pickup *scripts*.

What's a script?

It's a text, with stage directions, that if carried out properly achieves a consistent effect wherever and whenever it's performed.

The idea of a script may be off-putting to many people. To some men, it seems insincere: they'd rather just be naturally seductive without having to resort to canned material. To some women, it seems fake: they've learned through experience to sort through potential suitors and quickly make a binary decision about each—*yes* or *no*. And the idea that men can fake the qualities that elicit a *yes* short-circuits this entire evaluation system.

So I urge you not to use the following scripts. Instead, make up openers, stories, and gambits that are true and interesting to you.

However, before coming up with your own material, you might want to look these over anyway. Because they work. And they helped transform me from a guy who was too scared to speak to women into a guy who had experiences that, even in his wildest fantasies, he never imagined were possible.

One of the most interesting things about the so-called seduction community is that it functions like an international laboratory. Every routine that follows is something I've done myself scores of times with success. And only afterward did I share it in the seduction forums, where tens of thousands of men all over the world tried them out within days. Their feedback quickly made it possible to identify which routines were universally effective.

Even as I became more successful and natural in my approaches—able to say pretty much whatever was on my mind—I still found these routines valuable as fulcrums to move to the next stage of an interaction before the conversation hit a lull. Other times, even when I was in a relationship, they proved useful simply to liven up a boring dinner party or to win over a business contact.

The best of the best are included here (all in updated, improved versions)—except, of course, for the routines that were included verbatim in *The Game* (such as the Jealous Girlfriend Opener, the Best Friends Test, the Cube, the Evolution Phase Shift, and the Dual Induction Massage).

Because I've shared variations of these routines over the years, I wanted to make sure they hadn't become too widespread before publishing them. So I put together a street team of students from around the world to test the material. Before each routine, I've listed the results of their approaches, including the *difficulty level* (how easy or challenging the material was), the *saturation level* (how often students were accused of using game), and the *success rate* (how often the routine achieved its intended effect).

One interesting thing to note is that, even in the rare cases where the material was recognized, most students were still successful as long as they didn't get flustered, didn't lie about it, bonded over this unexpected commonality, and comfortably continued to talk to the women they'd approached.

Remember that there are no magic powers in the words on the following pages. What makes them succeed is your delivery. Reciting them like a memorized grocery list won't lead to a rich and varied social life. Instead, understand why these routines work before using them, be genuinely curious about the questions you're asking, and share the material more for your own amusement than to attain a desired response.

Like a comedian or an actor, you should connect with your audience members, making it seem as if your words are only for them in that moment. However, unlike traditional theater, the key to success in social interactions is improvisation: be willing to accept interruptions and unexpected reactions and run with them, rather than trying to finish the routine exactly as written. And, by all means, feel free to modify each of these scripts to fit your own personality and interests.

Finally, make sure you reread the section of *The Stylelife Challenge* associated with each type of routine before attempting it for the first time, so that you learn the timing and subtleties of the delivery. When using the openers, for example, make sure you add your time constraint and root. And when you start generating your own routines that work just as consistently, don't forget to share them with your friends, fellow challengers, and favorite authors.

YOUR FINAL WARNING:

ROUTINES ARE THE
SPAWN OF THE DEVIL.

THEY CAN LEAD TO DEVIANT
SEX AND TEENAGE PREGNANCY.

DO NOT USE THEM . . .

. . . UNLESS YOU REALLY, REALLY HAVE TO.

THE TWO-PART KISS OPENER

Type of Routine: Opener
Difficulty Level: 3/10
Success Rate: 91%
Saturation: 1.6%*
Comments: "The reactions are always good. This is my usual default opener. It's a great one and usually always leads to interesting follow-up questions." —GRANDMASTERFLEX
Origin: While writing *The Game*, I accompanied Courtney Love to an awards show. At one of the after-parties, her boyfriend was upset at her because, from time to time, she made out with women. She said she didn't consider that cheating. He said he did. So we decided to take a poll in the room.

YOU: Hey, guys, we're having a little debate and need a quick take on something.
GROUP: What's that?
YOU: If a guy is dating a girl, and she goes out to a bar with her friends one night and makes out with a guy just for fun, do you consider that cheating?
GROUP: Yeah, it's cheating.

* The statistics reflect the results of up to 1000 individual tests of each routine.

YOU: Okay, that makes sense. So here's the real question. And I'll tell you why I'm asking in a second. If she goes out and gets drunk and makes out with a *girl* for fun, is it cheating?

GROUP: [*The responses will vary, but if any guys say "no," you can call attention to their double standard (but with a smile—always).*]

YOU: Okay. Interesting. The reason I'm asking is because my friend over there has been dating this girl. And she likes to go out and get drunk and make out with girls. Now, some guys might be into that, but it upsets him and he thinks it's cheating. She says it isn't. So we thought we'd get an outside perspective on the situation.

GROUP: Well, it really depends on . . .

THE LOVE VERSUS IN-LOVE OPENER

Type of Routine: Opener
Difficulty Level: 3/10
Success Rate: 88.8%
Saturation: 1.5%
Comments: "Cool opener. I opened six sets and it worked every time. Most of the girls said they used the same line because they were just not into the guy."
—LosDog
Origin: I met a girl online who sent me topless pictures of herself, then came straight to Project Hollywood (the house where I lived while writing *The Game*) to have sex. For some reason, I'd noticed that many women who just want sex, but have little time for formalities like phone calls, meals, or spending the night, are often cheating on a husband or boyfriend. So as we lay in bed together afterward, I asked her if she was dating anyone. She replied that she was married. "I love him," she confessed, "but I'm not *in* love with him." It seemed like a small distinction, but for her those two letters were important enough to make all the difference between faithfulness and infidelity.

YOU: Hey there, my friend and I need a female point of view on something: What do you think the difference is between love and being *in* love? She wants to know because her boyfriend just broke up with her. He told her he loves her, but he's not *in* love with her. What's weird is that some girl made the exact same speech to another friend of hers recently. So we've been trying to figure out what the difference is exactly.

GROUP: I think the difference is . . .

YOU: Yeah, I guess that makes sense. Because I can give my best friend like a bear hug and say, "I love you, man." But if I say, "I'm *in* love with you," he'll probably freak out and punch me.

THE ALBINO GARY COLEMAN OPENER

Type of Routine: Opener
Difficulty Level: 5/10
Success Rate: 85.5%
Saturation: 1.8%
Comments: "The routine was very successful when used properly. It's more difficult to use with a group that includes both men and women. A good way to do this is to first ask the men what they think, then ask the women and compare both answers." —Major
Origin: One day, a pickup artist known as Swinggcat and I were at a bar, waiting for a friend of his to arrive. His friend is a confident, gnomish-looking guy who'd just gotten out of a relationship, so to kill time we decided to find out if any women there were interested in dating him. Not only did this work as an effective opener that night, but, in the process, we discovered the qualities each woman found most attractive in men—and in what order of importance.

YOU: Hey, I'm about to meet a friend here, but before he arrives I need some quick advice for him that you'll probably know.

HER: What's that?

YOU: His girlfriend recently broke up with him, and tonight is his first night out after recovering from it. He wanted us to give him some advice on meeting women, but you seemed like more of an expert. So out of curiosity, while we wait for him, what do you think is the number one thing that women look for in a guy?

HER: Sense of humor [*or whatever*].

YOU: Okay, sadly, he has no sense of humor [*or whatever*]. Is there anything else that women look for in a guy?

HER: Maybe if he's really rich [*or whatever*].

YOU: Well, he's not really rich [*or whatever*]. In fact, he just lost his job at Taco Bell.

HER: Well, if he's not funny and he doesn't have a job, I wouldn't date him.

YOU: What if he's the best-looking guy you've seen in your life? Or he's the smartest person on the planet? Or he's amazing in bed? There has to be something else.

HER: I guess if he was really intelligent [*or whatever*], and I could learn a lot from him.

YOU: Actually, he's not that intelligent [*or whatever*] either. Do you know Gary Coleman from that show *Diff'rent Strokes*? [*If you don't think she'll recognize the name, feel free to use a different pint-size actor like Mini-Me from the Austin Powers movies.*] He's five foot one and looks like an albino Gary Coleman, except he's not funny.

HER: [*laughs*] I don't think I'd ever date him.

YOU: That's okay. You're not his type anyway. [*A small percentage of people may be insulted if you say this without a teasing smile. If this happens, recover with:*] Just kidding. Thanks for the awesome advice. You should totally have your own radio talk show or something.

THE SPELLS OPENER

Type of Routine: Opener
Difficulty Level: 5/10
Success Rate: 88.6%
Saturation: 4.4%
Comments: "I had a hard time remembering the important tidbits of this opener, so I did actually struggle the first six times. But it kept on working despite my uneven delivery, so it's pretty damn good. And it transitions well in any number of directions, so it was easier to lead smoothly out of the opener into a value demonstration." —WINGDING
Origin: When I started learning the game, I signed up for the first-ever workshop that a pickup artist named Mystery was offering. He instructed us to approach women by asking, "Do you think spells work?" Since he was a magician, it was easy for him to follow up the line by making her drink levitate. But I, on the other hand, had no magic to follow-up with. Fortunately, I remembered that when I was writing a book with the guitarist Dave Navarro, an actress once came over to his house and left an attraction spell under the cushions of his couch. Two weeks later, she became his girlfriend. So I used this story to turn Mystery's random question into conversation.

YOU: Quick question: Do you think spells work—like magic spells? I know it's an unusual question, but I'll tell you why I'm asking in a second.

HER: I don't know. [*If she doesn't answer, just pause and then continue with the story.*]

YOU: Okay. You see my friend over there—he just moved here. And he met a girl at a club. He wasn't interested in her sexually, because she wasn't really his type.

HER: Sure he wasn't.

YOU: No, really. Anyway, so, she came over to his house a couple nights later to watch a movie and nothing happened.

HER: [*rolls eyes*]

YOU: Really, nothing happened. Anyway, after she left, he was cleaning, and he lifted up his sofa cushion and found a corroded metal ring underneath with a scroll inside and, um, some feathers wrapped around it. So he unrolled the scroll, and called me and said there were some weird, undecipherable letters on it. I told him it sounded like a magic spell. So he decided to bring it to this occult-and-spirituality-type store where they sell like vials of what they claim is dragon blood. And this woman there said it was an attraction spell.

HER: No way.

YOU: Yeah, and now the strange thing is, he's suddenly finding himself attracted to her. Like he can't get her out of his mind. Sometimes he'll be walking down the street, and an image of her surrounded by flowers will just pop into his head. So do you think it's the spell or it's just psychological?

HER: I think . . .

THE CAT PEOPLE COLD READ

Type of Routine: Opener/Value Demonstration
Difficulty Level: 3/10
Success Rate: 87.5%
Saturation: 2.1%
Comments:—"Responses were varied but enthusiastic, though everyone loved the part of the routine about dogs and babies. Distance seemed to play a factor when asking women to come open another group with you afterward. However, simply making it a smaller distance seemed to solve the problem." —Metz
Origin: Over dinner one night, the Stylelife coaches and I were reviewing the topics women seem to enjoy discussing most. One was relationships, another was spirituality, and a third was animals. We realized that there were openers about relationships (like the love versus in-love opener) and spirituality (like the spells opener), but none about animals. So we decided to play a game that night and try to guess what kind of pet women we met had at home. It eventually evolved into the following opener. Feel free to make up your own pet-owner theories based on your experiences with this routine. And don't forget to incorporate the cold-reading tips you learned on Day 15 of the Stylelife Challenge.

YOU: I have to ask, out of curiosity, which do you like better, cats or dogs? [*After performing this routine a few times, you should be able to open by guessing with some degree of accuracy whether each person in the group is a dog or cat person.*]

HER: I'm a cat [*or dog*] person. I actually have two at home.

YOU: That's so funny. My friend was just telling me he could figure out whether someone was a cat or dog person based on their personality. At first I didn't believe him, but he says cat people tend to be more assertive and have stronger personalities and convictions, which is why they get a pet that's more feminine. He said it's a yin-yang thing. Dog people tend to be more shy and gentle, which is why they want a more masculine animal around to balance them. It kind of made sense, so I thought I'd see if he was right.

After the opener, if there's someone else you want to meet nearby or you want to continue the interaction with her, you can follow with:

YOU: Before I go, we should quickly test it out. [*Pause.*] What do you think those people are?

HER: I'd say the girl over there is a cat person, and the one next to her is a dog person.

YOU: Cool. Let's go find out.

Approach the new group (either alone or with her) and repeat the opener from the top. You can return to the group you originally approached, continue to meet new people this way, or transition into the following afterward:

YOU: He also has another interesting theory. He says that when a woman in her twenties or thirties gets a dog, it means she's ready to have children. I guess it's because she wants to take care of something.

HER: Yeah, that's so true . . .

YOU: They also say that the way a guy treats his pet is the way he's going to treat his children.

THE FACEBOOK STALKER OPENER

Type of Routine: Opener
Difficulty Level: 3/10
Success Rate: 84.5%
Saturation: 0%
Comments: "I had one girl say she did exactly what the opener describes. I refused to give her my number and told her she'd have to stalk me on Facebook. I got her friend request the next morning." —CloverThief
Origin: Unlike the other openers, this comes from one of the Stylelife coaches, Stephen Grosch. It's included because it's something he uses to start conversations with bartenders and waitresses. What's interesting about the story is that it portrays him as the type of guy who's generous with his friends and whom waitresses find attractive. And, like all the other openers here, it's true. Ultimately, Stephen invited the waitress in the story to his house and they slept together, even though she had a boyfriend. Evidently, she loved her boyfriend, but wasn't *in* love with him.

YOU: Real quick, I need a professional opinion on something. The other night, a bunch of friends and I went to this amazing restaurant. Our waitress seemed cool as hell, and we were really hitting it off. At the end of the night, I was very close to getting her number. But then I decided not to because I thought she might have been flirting just to get a bigger tip. You know how waitresses will do that. Some will actually touch male customers on the shoulder when they talk to them because supposedly that makes them tip better.

HER: No way.

YOU: Yeah. So anyway, when we paid, since it was my turn to pick up the meal, I paid for it with a credit card. Then we left, and I figured that next time I'm in there, I'll see how it goes and then maybe get her number. Well two days later, to my surprise, I get a message from her on Facebook [*feel free to substitute your social networking site of choice*] saying how cool we all were and how much she'd like to meet me. Apparently, she copied my name off my credit card and looked me up!

HER: Oh my God.

YOU: I haven't responded yet, and I'm not sure if I will. But do you think that's cool of her to do or kind of creepy?

HER: I guess it's kind of cool and flattering.

YOU: So if you were eating out, and the waiter Facebooked you afterward, you'd think it's cool?

HER: Um, that would actually be really creepy.

YOU: So it's creepy if a guy does it, but cool if a girl does it? Isn't that kind of a double standard? [*If you're talking to a waitress or a bartender, add*:] And by the way, don't get any ideas: I'm paying in cash.

OPENER GRAB BAG

Type of Routine: Opener
Difficulty Level: 2/10 (average)
Success Rate: 88.5% (average)
Saturation: 1.5% (average)
Comments: "The opener about hair color was the easiest to deliver. It was quick and successfully opened low- to medium-energy sets. A root is crucial. My root: 'I just asked my buddy whether he preferred blondes, brunettes, or redheads. And he responded, 'I like ravens.'" —Jadebelly
Origin: Many openers involve asking someone for an opinion on a lengthy issue. However, there are many other ways to start an interesting, spontaneous conversation without appearing to hit on anyone. These can be especially useful in a noisy environment, like a club, where it's more difficult to tell a long, involved story. Here are a few quick additional openers.

- See that guy over there? He just told me he knows kung fu. Why exactly do you think he would say that to me out of the blue?
- See that girl over there? She just told me she's a white witch. What does that mean?
- Who would you trust more, an older woman or a younger woman?
- If a woman with brown hair is a brunette, a woman with red hair is a redhead, and a woman with yellowish hair is a blonde, what's the word for someone with black hair?

- What's the name for that dance people do in the Cirque du Soleil where they're hanging from those red ribbons? I want to learn that but can't figure out the name so I can Google it.
- Can you hold on to this for a sec [*hand her a glass or camera or phone*]? Thanks. [*Pause.*] A friend of mine taught me that the best way to butt into a conversation was to give someone something to hold. And I wanted to test it out.

NAME MNEMONICS

Type of Routine: Value Demonstration
Difficulty Level: 3/10
Success Rate: 96.9%
Saturation: 0%
Comments: "This is a great conversation piece. People really got into it, and had a great laugh the more ridiculous the mental image was." —Farmer
Origin: One of the routines Mystery and I used to teach at workshops was called the Peg System, which uses mnemonic devices to make it appear as if one has a perfect memory. One day, at a house party, I was demonstrating the peg system to a supermodel, and she said that her ex-boyfriend was a magician who used to do the same thing. She recommended a Website on mnemonics, from which I learned the following technique. I also searched online to find out if her ex-boyfriend was a pickup artist. It turned out he was David Blaine.

HER: What's your name?
YOU: I'm [*name*].
HER: I'm Hilary. This is Donna. And that's Tony.
YOU: Okay . . . Hilary . . . Donna . . . Tony [*pausing between each name and looking at each person*]. You know, I used to be really bad with names.
HER: Oh my God, I'm so bad with names.

YOU: But you don't have to be. You're probably just doing it wrong. Like, for example, I used to repeat names in my head and then, ten seconds later, forget them. Here, I'll show you super-quick how to always remember names, and then I really have to go.

HER: Okay.

YOU: It's great for meetings and things. All you have to do when introduced to someone is make a picture in your head. So if you're Hilary, I just imagine you but with Hillary Clinton's face. No offense. And for you, Donna, I just picture like the dawn, and the sun rising over your head. And for Tony, I see you on the front of a box of Frosted Flakes like Tony the Tiger. Here, try it.

Originally, I'd follow this up by having the group memorize my name. But it came off as slightly cheesy, like I was trying to trick them into remembering me. So, instead, grab a friend—or a stranger in the club (particularly another person you want to meet)—and teach the group how to memorize that person's first, middle, and last name. Let's say the person's name is Thomas Scott McKenzie. As you're doing this, remind them:

YOU: You can't just think about the visual image. You actually have to picture it in your mind. So try to imagine him wearing a Thomas Jefferson wig [*pause to give them time to visualize it*], but with the face of Scott Baio underneath it [*pause*] and, for McKenzie, say a big yellow McDonald's M in his hands with a Ken doll grinning out of the middle. So you have Thomas Scott McKenzie.

Feel free to substitute the references for anything you think the group will relate to. When in doubt, just ask if they know someone else with the same name and have them picture that person's face.

What's most fun about the routine is that later in the night (especially when Thomas Scott McKenzie passes by) or even on the phone the next day, you can quickly test them and see if they still remember the name. If you taught this correctly, after a little struggling, they'll always get it. If the routine still holds their interest, at some point you can also explain how it works:

YOU: Most people try to repeat someone's name in their head to memorize it, but they always forget it. This is because, when you're a baby, you don't have language. You have images, and that's how you interpret and understand the world. So making and storing pictures is the best way to commit anything to memory, because that's what we've been doing all our lives.

BLOOD-STROLOGY

Type of Routine: Value Demonstration
Difficulty Level: 6/10
Success Rate: 88.5%
Saturation: 0%
Comments: "This worked best when I had good rapport with the group and smoothly transitioned into the routine. It also worked on two sarcastic girls. When I told one she was probably an O type, then explained it, she agreed and actually said she'd get a test to find out." —Shure
Origin: While I was taking medical classes for the book *Emergency*, I learned which blood types were compatible in transfusions. And it made me wonder: If people put so much stock in astrology and how the movement of celestial bodies affects our behavior and personality, shouldn't they pay at least as much attention to the blood flowing through our veins and the effect that must have on us? So that night, while talking with George Rockwell, one of the Stylelife coaches, we decided to look into whether there was a science of personality reading and relationship compatibility based on blood type. The next day, he found a Japanese system of hemo-divination. So we made some modifications and ended up with this.

YOU: So I was with a friend from Japan the other day, and he was saying that people's personality—and the people they're compatible with—can be determined by their blood type. It's like their equivalent of astrology. Except he claimed it was more accurate, because astrology is about things billions of miles in the sky, but our blood runs through our body so it has a stronger effect. So far I've found this stuff to be surprisingly accurate. Do you know your blood type by any chance?

If she doesn't know her blood type:

YOU: Yeah, it would be good to know. I'm curious how accurate this whole thing is. I'll tell you what: let's try to figure out which personality profile fits you the best, and you can find out later if it matches your blood type.

Use the following descriptions to guess her blood type. If she actually checks with her parents or doctor, you've definitely made an impression.

If she knows her blood type, then proceed with the corresponding analysis below. Feel free to embellish it with your cold-reading skills. You might want to copy these descriptions onto your phone or a piece of paper, so you can refer to them if necessary. Just explain that you wrote down the information your friend gave you so you could remember it.

Note that the blood type compatibilities below apply only to interactions, not transfusions, which require more than a routine to perform successfully.

KEY

Type A: The Farmer. People with blood type A are known for being able to stay calm under pressure. They are hard workers who like to keep the peace and live comfortably, which can lead to strong relationships with the right person. However, they sometimes feel like outcasts, which tends to give them amazing artistic talents but also makes them somewhat shy and sensitive. They secretly crave success and are known to be perfectionists, though they can occasionally be stubborn and overly cautious. Type A is most compatible with types A and AB.

Type B: The Hunter. Type Bs are the most dependable of the blood types. They can be counted on to finish any project they start. They're good at following directions but prefer to find their own way to complete a given task. They tend to have one-track minds, and usually focus on what they're working on at that moment to the exclusion of everything else. They can seem cold, because they tend to stick to logic rather than emotion when dealing with people. They are often perceived as individualists and can sometimes appear selfish. Type B is most compatible with types B and AB.

Type AB: The Humanist. Type ABs tend to be passionate lovers but are also known for their somewhat unpredictable, dualistic nature: hot and cold, timid and confident, the life of the party and the shyest person you know. They tend to be easily overwhelmed by responsibility. AB types are known for being trustworthy and honest, but also generally have a dislike of custom and conformity. Type AB is compatible with all other blood types.

Type O: The Warrior. Type Os are known for being energetic, social, and ambitious. They follow their passions and tend to set trends, but when something doesn't interest them, they can get flaky. They're easy to fall in love with, but also dangerous for this reason. They love attention, and generally listen well to others. They tend to say what's on their mind (sometimes without thinking) and are generally confident, though they can also at times be jealous. Type O is most compatible with types O and AB.

TIME OUT

YOU: HEY, NEIL, THESE ROUTINES DON'T WORK.

ME: OF COURSE THEY WORK. I'VE DONE THEM HUNDREDS OF TIMES, AS HAVE THOUSANDS OF GUYS ALL OVER THE WORLD.

YOU: WELL, THEY'RE NOT WORKING FOR ME.

ME: AH, THAT'S DIFFERENT THEN. HAVE YOU COMPLETED THE STYLELIFE CHALLENGE?

YOU: [*SHEEPISHLY*] NO.

ME: I THINK I SEE THE PROBLEM. GO COMPLETE THE STYLELIFE CHALLENGE FIRST, THEN GET BACK TO THESE.

YOU: DO I HAVE TO? IT GOT HARD ON DAY 8, SO I STOPPED.

ME: YES. IN ORDER FOR THESE ROUTINES TO BE EFFECTIVE, YOU MUST UNDERSTAND HOW TO DELIVER THEM, WHY THEY WORK, AND, MOST IMPORTANT, AT WHAT POINT IN AN INTERACTION THEY SHOULD BE USED.

YOU: SO TIMING IS EVERYTHING?

ME: TIMING, TONALITY, BODY LANGUAGE, CONGRUENCE, CALIBRATION, CURIOSITY, SPONTANEITY, ATTITUDE, INTUITION, EXPERIENCE. EVERYTHING IS EVERYTHING. IT ALL COUNTS. THIS IS NOT JUST ABOUT WOMEN. IT'S ABOUT YOU, AND MAKING YOURSELF A BETTER PERSON. SO THE ROUTINES ONLY WORK IF YOU WORK.

THE FIVE QUESTIONS BET

Type of Routine: Value Demonstration/Playful Game
Difficulty Level: 2/10
Success Rate: 95.7%
Saturation: 0%
Comments: "This is one of the easiest bar bets to inject fun into the night and raise the energy level. With different girls, I did the bet for a phone number, a free drink, a truth or dare question, and a kiss—and got all of them." —20spot
Origin: While trying to improve my game, I spent a weekend with a hustler friend learning dozens of the betting games he uses to con bar patrons into buying him a drink. I noticed that even when people lose a drink, they almost always feel it was worth it for the entertainment. The following remains my favorite bar bet.

YOU: I'll tell you what: To decide who gets the first round of drinks [*alternately, you can bet a dollar, a cup of coffee, a hand massage, or anything small*], let's play a little game to make it fun. It's called the five questions bet.

HER: What's that?

YOU: I'm going to ask you five questions, and all you have to do is answer each one incorrectly. Just so you know, there's no trick question that doesn't have a wrong answer. It's very easy to win this, if you can just answer five questions in a row wrong.

HER: Hmm. Okay.

YOU: And make your answers as surreal as possible, so I know they're wrong. Fair?

HER: Sounds fair.

YOU: Do you want a practice question, or should we just start?

HER: Let's start.

YOU: Okay. What's your name?

HER: Bill. [*Some people will actually say their real name and lose on the first question. These are probably not the ones you'll want to consider for a long-term relationship.*]

YOU: Well done. What city are we in?

HER: Vatican City.

YOU: Great. Okay, now what's the name of this bar [*or café or mall or whatever*]?

HER: Wal-Mart.

YOU: Great. So . . . [*pause and switch the tonality of your voice; also look down, touch your face, and act a little confused*] how many questions was that so far?

HER: [*If she answers "three," then you win. If she figures out what you're doing and answers with another number, then proceed to the fifth question.*]

YOU: [*impressed and shocked*] Oh my God, you got me! Have you played this game before?

HER: No!

YOU: Ha—I got you! [*Pause.*] Thanks for the drink.

HER: Oh God, I can't believe I did that.

YOU: Here, to make you feel better about losing, I'll teach you how it works so you can win drinks from your friends. On the fourth question, always ask, "How many was that so far?" That question is designed to trip up someone who's helpful by nature. And the fifth question, "Have you played this before?" is designed to trip up someone who's proud or egotistical. So between those two questions, you basically have most people covered. And you're obviously not a helpful person. [*Or alternatively, "And you're obviously a helpful person."*] Good to know.

THE AMAZING TABLE PSYCHIC

Type of Routine: Value Demonstration/Magic
Difficulty Level: 5/10
Success Rate: 81.3%
Saturation: 0%
Comments: "This can be used as an opener if you choose to. Simply say, 'Hey guys, can I borrow you a sec? I'm trying to show my friend this thing because he doesn't believe it works.' This usually started some playful touching afterward as the girl kept trying to find out how I did it." —Symphonie
Origin: I was at Jerry's Diner in Los Angeles teaching a friend some of these value demonstrations. As we were talking, the manager, Mike, came by, and we started discussing game. After I explained the basics to him, he said he wanted to show us something he had learned in high school. He whispered a few words to my friend, and then this is what he did.

THE EFFECT

YOU: You know what's interesting? People leave energy behind wherever they go. Like, if you touch something, you leave a sort of psychic imprint.

HER: Really?

YOU: I'll give you an example. I'm going to gather some items on the table. [*As you talk, arrange six different items—such as a cell phone, keys, a crumpled*

> *napkin, and so on—in three rows of two on the table. The arrangement should look like the six on a die.*] When I leave the room, touch any object and I'll be able to tell which one you chose by feeling the energy on it. In fact, if you want, you don't even have to touch it. Just hold your hand briefly above the object.

Leave the room, and make sure it's clear to the woman that you can't see the table. When you return, hold your hand above each object to feel the energy. When you've built just enough suspense, make your selection.

YOU: Was it this object?
HER: Yeah! How did you do that?
YOU: [*pause, then reluctantly*] Can you keep a secret?
HER: Yes.
YOU: [*breaking into a smile*] So can I.

HOW IT'S DONE

Your wingman should have a drink and a cocktail napkin in front of him before the routine begins. When you leave the room, he observes which object the woman selects. At some point before you return to the table, he should take a sip of his drink and then place the glass on the edge or corner of the napkin that corresponds to the placement of the object she selected.

Remember that the objects are laid out in three rows of two. So if she selects the object in the upper-left corner of the group, your wingman places his drink in the upper-left corner of the napkin. If she selects the object in the second row, on the right side, your wingman places his drink in the center of the right edge of the napkin. And so on.

In my experience, no one's ever figured this out. But don't worry if she happens to know how you did it, she catches you, or you mess up. The point of this routine is not to convince women you're psychic, but simply to avoid mundane small talk and have fun. As long as everyone's smiling or laughing, the routine is succeeding.

THE LYING GAME

Type of Routine: Value Demonstration
Difficulty Level: 5/10
Success Rate: 75%
Saturation: 3.4%
Comments: "Although on paper this looks complicated, it went really well. Watching Derren Brown videos on YouTube, as well as searching Google for NLP and eye movement articles, gave me a greater understanding of the routine and more things to talk about. With the popularity of TV shows like *Lie to Me,* a couple people said they knew this, but I just told them this was more accurate." —Diamond
Origin: One of the hundred-plus books I read while studying social dynamics was *Never Be Lied to Again* by David Lieberman. At the time, I was interviewing the singer Carly Simon for the *New York Times,* and during the conversation I suggested she read the book. A week later, she called and said she'd bought ten copies. "So you really liked it?" I asked. She replied: "I didn't read it. I just bought a bunch of copies and gave them to all my friends, so they'd be too scared to lie to me." That conversation set off an exploration into more clever and subtle ways to spot lies, leading to this routine.

YOU: You know, there's a way you can always figure out whether or not someone is telling the truth. It's good to know, because if your boyfriend comes home late one night and says, "I was out bowling with the guys," you can bust him on the spot if he's lying.

THEM: I'd love to know that.

YOU: Okay, we'll need one of you to be the liar then. Hmm, you seem right for that. So here's what we're going to do . . .

Whisper the following to the rest of the group, so that the liar doesn't hear.

YOU: Watch for a change in her eye movements as she's answering each question.

To the subject:

YOU: Okay, do you have a brother or a sister?

LIAR: [*Answers the question.*]

YOU: Okay, choose one of them. [*If she doesn't have a sibling, have her choose her car or her bedroom.*] So for this test, you're going to think of five facts about your brother. Don't say them out loud, just think them. But don't think about each fact until I tell you to. I'll say "one," then think the first fact. When I say "two," think the second fact. And so on. Ready?

LIAR: Okay.

YOU: And make one of the five facts a lie. [*Give her this piece of information at the last minute, so that she doesn't think of the lie in advance.*] Okay, one [*pause to give her time to think*]. Two [*pause*]. Three [*pause*]. Four [*pause*]. Five [*pause*].

To the rest of the group:

YOU: Okay, which one do you think was a lie?

GROUP: Definitely number three.

YOU: I'd say either number one or number three. [*Consider giving two answers, just to increase the odds that you're correct.*]

HER: It's totally number one. How did you know?

YOU: When people remember things, they go to one part of their brain to access the information. But when they make things up, they go to a different part of their brain. And I could tell when you lied by looking for the variation in your eye movement, which meant you were getting information from the creative part of your brain instead of the place where you keep your memories.

If no one is able to determine which answer is a lie, continue with:

YOU: Wow, you're a good liar. I could never date you. Fortunately, there are two other ways to tell when people are lying. One is to see if they break eye contact. Usually, liars know that they need to keep eye contact while they're telling a lie, but they'll often immediately look away afterward. The other way is—here, I'll just show you quickly. Let's choose someone else this time. [*Select someone else in the group.*] Okay, tell me three things you did yesterday—say them out loud—and make one of them a lie.

When she says three things, choose each statement, starting with the one most likely to be a lie. Then ask her lots of questions about the details as fast as you can. For example, if she says she went swimming, ask, "Where did you go swimming? For how long? Describe the swimsuit you wore. Who was in the lane next to you?" Her answers to the questions don't matter. What you're looking for is a point where she stumbles on her words or hesitates or starts laughing or breaks into a guilty grin. That's when you'll catch her in the lie.

YOU: That's the third way. You just ask lots of rapid-fire questions, the way they do when you're going through Customs. The answers don't matter. Just keep firing off questions until they trip up. Like she just did.

THE FAT BASTARD CHALLENGE

Type of Routine: Playful Game
Difficulty Level: 3/10
Success Rate: 87.5%
Saturation: 0%
Comments: "I didn't receive as many negative responses as I was initially anticipating. The routine seemed to hook well and everyone had something to say about it. I was weary of the gun to the head option, but I didn't notice any negative feelings toward it. I was also surprised by how many girls accepted the offer at one million dollars. One guy said he'd have sex with Fat Bastard for a happy meal." —Pendragon
Origin: I was having dinner with a friend named Kendra, and one of my favorite things to do when I'm bored (sorry, Kendra) is to ask people—in a fun, nonjudgmental way—what they would do for money. So for some reason, I ended up asking how much money she'd require in order to agree to sleep with Fat Bastard from the Austin Powers movies. I was surprised by her answer. Not only would she not do it for a hundred million dollars, but she said she'd rather die than sleep with him.

YOU: I'm curious about something. Do you know that character Fat Bastard from the Austin Powers movies?
HER: Yes.

YOU: If someone came up to you and said they'd give you a million dollars if you slept with a guy who looked like Fat Bastard but ten times worse, would you do it?

HER: No.

YOU: Then would you do it for ten million dollars? Just once?

HER: No way.

YOU: What if it just lasted half an hour? You could think about other things and go to that special place in your mind. Then afterward, you'd have ten million dollars tax-free.

HER: Maybe I could get him really drunk, and make him think we slept together or trick him somehow.

YOU: No, you have to sleep with him.

HER: Then no.

YOU: Okay, how about a hundred million dollars? Think about it: one hundred million dollars in your bank account.

HER: I still wouldn't.

YOU: Come on. Even I'd sleep with Fat Bastard for a hundred million dollars.

HER: I wouldn't do it.

YOU: Really? What if no one else knew or would ever know, not even me?

HER: Um . . . I don't know. I don't think I would. Maybe. I don't know.

Tellingly, half the women who still say "no" at this point will falter when asked the above question.

YOU: Okay, last question and then we'll talk about something boring, like the weather. What if someone put a gun to your head and told you, "Either you have sex with Fat Bastard or you die." Which would you choose?

HER: I'd try to escape.

YOU: Hey, now, this is a hypothetical question. You don't have other options. Either you sleep with Fat Bastard or you die. What would you do?

HER: [*Answer.*]

YOU: Interesting. What's funny is that every guy I asked that question said he'd rather have sex with Fat Bastard than die, but half the women said they'd rather die. Why do you think that is?

STYLE'S EV

Type of Routine: Emotional Connection/Value Demonstration
Difficulty Level: 5/10
Success Rate: 89.7%
Saturation: 0%
Comments: "While I found this a little more difficult to do than other routines, it was worth it. The first time I used it, I got a phone number. The woman really enjoyed the exercise and said she felt like she had learned something really cool about herself." —Erehwon
Origin: When I began learning Neuro-Linguistic Programming (or NLP), I was told to find out a woman's core value and trance words, and then to use her exact language to convince her to do things she might otherwise resist. I soon discovered, however, that it was more effective and ethical to practice NLP overtly instead of sneakily. So I turned the process that some therapists use to learn people's core motivating beliefs (known as eliciting values or EV) into a discussion that would teach them—and me—something about themselves. Today, I use this routine not just with women I meet but in almost every magazine interview I do. It's a great way to get to know someone very quickly.

YOU: Hey, as long as we're talking, let's do something interesting. Someone just did this with me recently. It's a great, quick way to get to know someone. In fact, a lot of people don't even know this about themselves.

HER: What's that?

YOU: It's just three questions. It's easy, and it'll tell you what really drives and motivates you in life.

HER: That would be cool. What's the first question?

YOU: The first one is: If you had to choose one thing you need to have in your life in order to feel like life is worthwhile, what would it be? [*If this question is too abstract for her, ask instead, "Name something you really enjoy doing."*]

HER: [*names a worthwhile thing*]

YOU: Okay, if you have [*worthwhile thing*] in your life, what kinds of things does that allow you to do or experience? [*If the second question is also too abstruse, ask instead, "Describe your perfect experience of (worthwhile thing). Either the best time you had doing it or your ideal scenario of (worthwhile thing)."*]

HER: [*names various things*]

YOU: Okay, imagine a time in the future or even now when you have [*worthwhile thing*] in your life. And this enables you to do [*various things*]. [*Paint an ideal picture here but make sure you use the exact words she's spoken, because they mean something special to her beyond just their dictionary definitions.*] How would that make you feel inside?

HER: I don't know. I guess good.

YOU: Go inside a little more. Just a minute ago, you smiled as you were imagining it. What was that feeling you got inside? Right there.

HER: [*names feeling*]

This will usually be a word like "fulfilled," "safe," "free," or "joy." Try to direct her toward a more specific word if she answers with a vague feeling instead like "good" or "happy"; if she's still unable to elaborate, tell her, "Well, maybe it's difficult to capture a feeling like that through language, but that feeling you get inside when you say that word—that is your core value."

YOU: Yes, that's it. [*Feeling*] is your core value. In other words, it's what really motivates you. Some people say they want to be an actor, and they think it's because they want to be famous. But the truth is, what they really want is to feel [*feeling*]. And it's funny, because when we were talking about imagining it earlier, you actually felt it for a second. It was really cool.

HER: Yeah, I did.

YOU: Awesome. We fulfilled your life goal in five minutes. You can die now [*pause*]. But seriously—and this is the real lesson—whenever you have to make an important life decision, whether it's about a job or a guy or a friend, just ask yourself if it brings you closer to that [*feeling*] feeling. If it does, then you should pursue it. If it doesn't, then you should move away from it.

HER: Wow. That's really interesting.

YOU: That'll be fifty dollars. I don't do this shit for free, you know.

THE SECRET SELF ROUTINE

Type of Routine: Emotional Connection
Difficulty Level: 5/10
Success Rate: 100%
Saturation: 0%
Comments: "I was actually surprised by how easy it was to have girls tell me what they didn't like about themselves. It's fun too. I think it's actually easier than Style's EV because they have less to think about. They all came out with some pretty random examples of how their characters looked, which made for some laughs." —Sandoval
Origin: One evening, I was at dinner with my girlfriend at the time and Billy Corgan from the Smashing Pumpkins. He walked us through an exercise he had done with one of his teachers, which helps people with their inner demons, issues, and insecurities. Since then, I've done it with hundreds of other people. It can be a powerful, transformative experience. However, not every woman will be willing to expose this vulnerable side of herself. So make sure you have a sufficient degree of trust, comfort, and rapport before starting this discussion. If she seems uneasy with the topic, just drop the routine and talk about something else.

YOU: You know, a lot of people try to repress the parts of themselves they don't like. But that never works. When you try to repress something, you're basically pushing it down on a spring. Eventually, it's going to release full force and take over your personality. It's interesting, because a friend recently did this psychological test with me, and it taught me that instead of denying the parts of yourself you don't like, there's a better way to handle them.

HER: What's that?

YOU: I'll tell you what. I'll quickly do it with you. It's just four questions. But for it to work, you have to be totally honest.

HER: Okay.

YOU: The first question is the toughest. What's the part of your personality that you like the least? This is the part of yourself that you don't like to show other people—your secret self—which maybe you sometimes even wish you could get rid of. [*If you're dealing with someone who's not self-aware or doesn't understand the concept, list a few common negative traits for her until she picks one or you assign her one that's most appropriate.*]

HER: [*says a negative trait*]

YOU: Okay, if you could give this part of you a name, what would it be? For example, a friend of mine said his problem was that he was too controlling, and he named this part of him Dexter.

HER: Okay, I'll call it [*says a name*].

YOU: Good. What does [*name*] look like? Describe her features and what she's wearing. For example, my friend said Dexter was a red baby, floating in the air with a pitchfork and a forked tail like a devil.

HER: [*gives a description*]

YOU: Okay. Now here's the key question. That part of ourselves that we don't like probably once had a purpose that no longer serves us. So if we give it a new purpose that's helpful to our lives, we don't have to repress it anymore. For example, when my friend did the exercise, he had to find a useful job for his controlling nature. And since he's an actor, he made Dexter his manager. So Dexter helps him rehearse, gets him to the set on time, critiques his performance, and drives him to make the right choices about his career. Another friend of mine had an anger problem, but now he uses that energy as his personal trainer in the gym to make him work out harder. So for [*name*], what job can you give her that would be constructive in your life rather than destructive?

HER: [*says a job*]

If she has trouble finding a job for her secret self, you may want to suggest some occupations or reframe her trait into something positive. Afterward, in order to avoid getting stuck in the "therapist" role, switch gears and say something like:

YOU: That's perfect. So [*name*] can be your [*job*], and help you with your life rather than hindering it. It's a pretty amazing exercise. I think we need to talk about something shallow now, though, like reality TV.

THE NANCY FRIDAY FANTASY ELICITATION

Type of Routine: Call to Action
Difficulty Level: 4/10
Success Rate: 100%
Saturation: 0%
Comments: "I love this. Having a girl tell you her sexual fantasies just minutes after meeting her is surreal. I ran this on an engaged woman who was due to be married in five weeks, and she basically offered herself to me on a plate!" —Tigs
Origin: For most of my life, I thought that sex was something a woman had to be tricked out of—usually with a wingman named Jack Daniel's or Grey Goose. So when I was learning the art of attraction, I hit a sticking point for several months. I could meet a woman and exchange phone numbers, but I couldn't work up the courage to turn talk into touch. My first step toward solving that problem was reading books that brainwashed me into accepting the truth: women not only want sex just as much as men, but also usually enjoy it far more. One of the first books I read was Nancy Friday's *My Secret Garden*. There comes a time in a conversation to start turning the subject safely toward sex without coming off as horny or desperate, and the book provided a perfect launching point for doing so (note: the story of the book's origins remains unverified).

YOU: Women are so much more fascinating than men. For example, there was this professor in the sixties who wrote a book and said that women were incapable of sexual fantasies.

HER: That's not—

YOU: I know, exactly. Obviously it's not true. So this woman named Nancy Friday wrote a book in response called *My Secret Garden*. And to disprove his theory, she interviewed hundreds of women about their sexual fantasies. Where men's fantasies are handed to them on a silver platter and encouraged, making most of them pretty much the same, women live this much more exciting and varied fantasy life. I think this is because women's sexuality is often repressed when they're young. If they see their dog or their father peeing, they ask, "What's that?" And they're told, "That's bad. Only bad girls talk about that." So their sexuality is held back, and eventually starts to flower in dynamic, wonderful ways.

HER: That's interesting.

YOU: Yeah, so Nancy Friday interviewed these women who were basically in relationships where they never even had oral sex and had sex only in the missionary position, and they had these wild fantasy lives. So she says that a woman's mind is like a house. And each room contains a different fantasy. There's the anonymous sex room. There's being with other women, being watched by an audience, being dominated, being a prostitute, or even transforming during sex into something or someone else. Obviously not every woman has all these rooms in her mind. Like, for example, when you're alone and thinking about something that gets you excited—and it doesn't have to be anything you've ever done or would ever do in real life—do you think about something that's in one of those rooms, or something completely different?

HER: I guess—

YOU: It's funny. A lot of people think things that get them excited but actually want them to remain just fantasies. Like, I dated someone whose fantasy was to be on stage strapped in stirrups in this mechanical device, while these robots had sex with her and an auditorium full of doctors in white lab coats watched. [*Pause.*] And, no, we never did end up doing that.

HER: [*describes her fantasy*]

YOU: That's interesting. It's amazing. Women can have all these different kinds of orgasms—vaginal, clitoral, blended, full-body—and usually they can have many of them, back to back. While most men only get this one little release that isn't nearly as pleasurable or intense. So you'd think it would be women who chase men for sex and not the opposite.

Rather than lingering on this subject, temporarily change to a nonsexual topic shortly after the conversation and allow her to think the ensuing thoughts while you're chattering away passionately about something completely unrelated.

THE SEVEN-MINUTE DATE

Type of Routine: Call to Action
Difficulty Level: 6/10
Success Rate: 65%
Saturation: 0%
Comments: "This routine takes a strong frame to pull off. I had to make the routine fun and playful in order for it to be effective, and it had to seem spontaneous. Separating the woman from her friends was the biggest problem, and I either needed a good wing to keep the group entertained or a female strong enough to brush off her friends." —DocDan
Origin: One of my favorite pickup artists to go out with while writing *The Game* was known as Maddash. Every night, we seemed to turn rooms full of strangers into best friends. Along the way, we'd improvise material like the Whole Room Destroyer (bonus routine alert), where we'd eliminate any competition in the room by telling women, when they were laughing and enjoying something we said, "Listen, I'll tell you what. Pick any other guy in the room, and I will personally walk up and introduce you to him. And I guarantee you that not one of them is as interesting as us." If they didn't pick anyone else, Maddash felt they were basically conceding that we were the most interesting people in the room. Below is another piece of nonsense we came up with that same night. It was based on a principle we'd read in the book *Influence*: that the best way to sell something is to lower the cost of commitment. So we decided to lower the cost of commitment of going on a date. As indicated by the lower success rate for this routine, calibration is key. Make sure the woman has some attraction to you first. And if she's reluctant to separate from her friends, just invite them along as chaperones.

YOU: You seem cool, and I can totally imagine wanting to hang out with you again. If I was one of those guys over there, I'd probably ask you out on a date. But dates are so awkward. Who wants to spend four hours sitting across the table from a stranger who you may end up not even liking?

HER: I know. I could tell you stories.

YOU: Hey, so instead of doing that, let's have our first date right now. We'll go on, say, a seven-minute date. [*Check watch.*] It's 9:50 right now. How about meeting at 9:52 at, say, that table right over there.

HER: Okay. See you there. Don't be late, now.

Meet her there. Time the date. Perhaps do one of the value demonstration routines from this book on the date.

YOU: Wow, it's 9:59 already. Time really flies with you. Thank you for a lovely evening. And don't get any ideas, because I don't kiss on the first date.

Shake her hand formally. Afterward, you may either end the date:

YOU: Listen, I think this is going too fast. It's not you. It's me. I'm just not ready for a serious relationship right now. I hope you'll understand and not take it too hard.

Or go on a second date, after which you can now kiss her good night.

YOU: Hey, listen, I had a really great time with you tonight. We should do this again sometime. Are you free in, say, two minutes?

THE QUADRUPLE HAND TEST

Type of Routine: Physical Connection
Difficulty Level: 3/10
Success Rate: 91.7%
Saturation: 0%
Comments: "Whenever I used this, it got a playful interaction going and ended in a really good make-out. If the girl didn't comply, I would turn around, grab her hand, and lightly smack it as a playful form of punishment. One girl gave me the pouty lip. Those pouty lips found my lips not long afterward." —Drewder
Origin: One of the biggest mistakes I used to make was simply lunging for a kiss when the time seemed right. Once I learned to escalate physical contact naturally toward kissing, I came up with the Evolution Phase Shift routine included in *The Game*. But eventually I became more comfortable with sexual tension, and I discovered that I could reach the same goal with fewer words and better body language.

When you sense that she would not only be comfortable but would appreciate more intimate contact, lead her to another area to meet a friend of yours, get a drink, see something interesting, or simply "make a lap." Remember that when making initial physical contact, never pull her toward you.

YOU: Here, come with me.

TEST ONE

As you get up, walk ahead of her and bring your right hand back to take her hand. Try to leave your hand flat, so that she grasps your hand first. If for any reason, she sees your hand and either doesn't take it or just lets her hand lay flat in yours without holding on, this generally means you haven't calibrated correctly and she isn't ready for physical contact yet. If she takes your hand, proceed to the next test.

TEST TWO

As you grasp her hand and lead her across the room, slowly release your grip but don't otherwise change your hand position. If she continues to hold your hand, proceed with the next part of the routine.

TEST THREE

Once you arrive at your destination, release her hand and don't attempt to touch her again for a few minutes. At some point, start talking to a third party—a friend, the DJ, a salesperson. If no one else is there, tell her you need to make a quick call to check a message about a party you're supposed to go to later.

While you're distracted by this other person (or your phone) and talking with your head turned away from her, casually take her hand in yours. Do this with the attitude that you're just reassuring her that you're still conscious of her presence, and can't wait to get away from this distraction so you two can continue talking. The less you look at her and the more your body is turned away from her, the more comfortable she'll be with your touch. If she's comfortable holding your hand and makes no motion to let go, then proceed with the fourth and final test.

TEST FOUR

As you're talking to the third party, rub your thumb reassuringly along the outside of her hand. If she massages your hand back, then she is ready to be kissed.

However, there is no need to kiss her right away. Continue holding hands for a minute or two, then let go. You should always be the first to break physical contact. Now that you know she's interested and the window to physical intimacy is open, you can wait anywhere up to fifteen minutes before transitioning into your first kiss.

STYLE'S KISS CLOSE

Type of Routine: Physical Connection
Difficulty Level: 5/10
Success Rate: 81.5%
Saturation: 3.6%
Comments: "I tried this five times with the quadruple hand test and three times without it. The one time it didn't work was without the hand test. I also had one girl recognize it. She mentioned that her ex had said the same thing during their first kiss. However, she still kissed me."—Prodigy Alpha
Origin: There's no routine that will make a woman kiss you if she doesn't want to already. The only point of a kissing routine is to bridge the gap into intimacy comfortably, without triggering her auto-pilot lip-deflection response. I first used this routine while doing a psychological exercise called the Cube with a dancer I had met. I still had ten minutes left in the routine, but knew that the window to kissing was about to close if I didn't act soon. Note that, contrary to popular lore, the first kiss works best in isolation in the middle of the date rather than at the end of the night.

As you're in the middle of telling any story or routine, start to falter and pause as you continue looking in her eyes, like you're being distracted.

YOU: . . . so then we just said . . . um . . . we'd have to lift it . . . out . . . Stop it.
HER: What?
YOU: Stop looking at me like that. You're distracting me.
HER: Like what?

Take the top of her head, and gently turn it away from you.

YOU: Okay, much better. Now where was I? Oh, yeah. So then we just said we'd . . . Fuck it. Come here.

At this point, if she's looking at you and smiling, you can lean in and kiss her. (Avoid leaning in all the way; she should do at least a little of the work.) If you're not sure if she's ready to be kissed yet, instead of saying, "Fuck it. Come here," continue with:

YOU: God, I'm trying so hard *not* to kiss you right now. Stop looking at me like that.

After saying this, watch her response. If she holds eye contact or looks down shyly, you may slowly move in for a first kiss.

If her body language closes off, or she starts to make an excuse or a condescending comment like "Aww, you're so sweet," then say, with a smile, before changing the subject:

YOU: Easy, now. I need trust, comfort, and connection first. I flirt a lot, but I don't put out. [*pause*] Despite what everyone else here will tell you.

THE LAST-MINUTE-TENSION ELEVATOR

Type of Routine: Physical Connection
Difficulty Level: 6/10
Success Rate: 82%
Saturation: 0%
Comments: "Although I was skeptical of this routine at first, when it came time to take off her bra and she said no, I agreed and went to play with my parakeets. Within five minutes, we were both back in bed. Within ten minutes, we were naked." —Bones
Origin: Though men tend to get anxious before approaching a girl, women tend to get more nervous when they're about to cross the point of sexual no-return. Most guys make the mistake of rushing toward the finish line, when, ironically, slowing down and letting go of a particular physical outcome will get them there quicker. This routine is a quick distillation of the knowledge on the subject I learned in the community, in psychology books, and from experience.

STAGE ONE: ESTABLISH TRUST

At some point, well before she's ready to be intimate with you, tell her something like the following:

YOU: A lot of women worry about how long to wait to sleep with someone. And whenever they ask me how many dates they should wait to have sex with someone they like, I just think, "You don't understand men at all."

HER: What do you mean?

YOU: I mean, some guys will wait as long as it takes to sleep with a girl. They'll hang in there for months just to "get some." What most women don't understand is that before the first date is over, the guy already knows if she's girlfriend material.

So the real truth is, you should never sleep with a guy until he knows your value. In other words, until you know he appreciates you for something more than sex—for who you are—don't sleep with him. But sometimes you can meet someone right away and just connect, and it's right. Whether you sleep with him that night or a week later, he's still going to feel the same about you. In my experience, the most passionate relationships usually begin passionately. Most normal guys don't actually judge girls for giving it up too quickly. They're glad to get that whole awkwardness over with so something more can develop.

HER: Maybe. I don't know.

ME: Really. I had this friend who was a bit of a player. And one night, he went out, met a woman, and had sex with her in the bathroom within fifteen minutes. And now they're married. If there's chemistry, there's chemistry. The other thing that's important, though, is you want to make sure you're with a guy who's discreet, who doesn't kiss and tell. You're not one of those people who goes and tells your friends everything afterward, or, like, Twitters all about it, are you?

HER: No.

YOU: That's a relief. I feel like it's called private life for a reason. It should be private.

STAGE TWO: BE THE FIRST TO STOP

You never want a woman to tell you to stop or slow down. To keep this from happening, suggest the idea before she does. So if you start to sense resistance or anxiety from her, stop kissing or remove your hands from whatever they're touching, and say something like the following:

YOU: Whoa, we should stop. This is going way too fast.
HER: Yeah.
YOU: Let's talk a little. Like, uh, have you ever ridden a horse before?

If she tells you a story, listen to it and talk a little, then start making out again. If she seems bored by the question or story, and wants to continue making out, just say:

YOU: Oh, fuck it. Come here.

You can do this two or three times before proceeding to the next stages, if necessary.

STAGE THREE: PACE HER REALITY

If you don't manage to raise the objection before she does, simply agree when she says that things are going too quickly. Remember, when you're in the heat of the moment, a logical argument will only make things worse. What she's responding to is a feeling. So slow down a little but keep the passion high, so that you're both agreeing that you're being "bad" while continuing to make out and turn each other on.

HER: We should stop.
YOU: You're right. We totally shouldn't be doing this right now.
HER: Oh God, I know.
YOU: We are so bad for making each other feel this good.

STAGE FOUR: FREEZE OUT

As a last resort, after using the other techniques, if she continues to stop you or pull your hand away, just give her space—more than she wants. Make sure you do this casually, without showing any sign of being upset or disappointed.

Sometimes it's enough just to create distance, and she'll reinitiate contact. But there are times when a more extreme takeaway may be necessary. In this case, switch on the light, turn off the music, and then check e-mail on your laptop, watch a boring TV show, or answer your cell phone and have a fantastic conversation with someone.

HER: Is everything okay?

YOU: No worries. I just really respect people's boundaries. I've always done things safely, maybe even to the extreme, probably because of the way I was raised. When someone says no, I assume they mean no and I just shut down. It's the way I am, and I think it makes sense. When I make love with someone, I want it to be amazing and something we'll be happy about the next day and will lead to better and better sex each time. And if the person I'm with doesn't feel like this is something she totally wants to do, and can really give herself permission to let go, then it's just not right. So it's really no problem.

After saying this, most often she'll sidle up to you and reinitiate contact. If not, you can work or watch TV for a little bit, then just say, "Come here," and continue again. Often, if you simply say, "Lift your arms" or "Take off your shirt," she'll be more comfortable doing so now.

STAGE FIVE: BE EXCELLENT

Despite all of the above, the best thing to do, besides making sure never to make her feel like a slut or like she's being used for your gratification, is to be really good in bed. Not like a porn star—those films are made to fulfill men's fantasies—but like a character from a romance novel, who intuitively understands both a woman's body and her feelings. If you can teach her something new about her own body and help her feel pleasure she's never experienced before, she will want to be with you just for the learning experience. Read books such as The Guide to Getting it On *or Mantak Chia's* The Multi-Orgasmic Couple, *and if you're able to demonstrate authority over a woman's body without ever seeming desperate or needy, she'll be the one seducing you.*

THE DOUBLE DATE THREESOME

Type of Routine: Physical Connection
Difficulty Level: 6/10
Success Rate: 100%
Saturation: 0%
Comments: "I had my first threesome ever thanks to this. And I'd never even been with the second girl before. When they came over, I had Nancy Friday's book on my desk and I started talking about different fantasies. Then I left the room. When I came back in, they were making out. I got on the bed and said, 'This isn't fair. Come here.' I kissed one girl, then the other girl. Ten minutes later, they were blowing me on the airbed and I was thinking, 'I hope this bed doesn't pop.' It was honestly the most amazing thing ever." —Hype
Origin: In *The Game*, I wrote about using the Dual Induction Massage to initiate a threesome. But once I realized how amenable many women are to a threesome if it seems to happen comfortably and spontaneously, I found I could be slightly more overt. So I started using the general methodology below. It's designed to bring together two women you're sleeping with on a nonexclusive basis.

1. When you're both alone, ask Woman #1, "Have you ever kissed a girl before?"
2. On a separate occasion, ask Woman #2 the same question.

3. If both have either had an experience they didn't regret with a woman or are curious about it, proceed with the rest of this routine.
4. Make sure you never bring up the subject of wanting a threesome with either woman at any time. After asking the question above, avoid mentioning it again.
5. Make plans with both women for the same night. Tell each separately, "Let's meet at my house, then maybe we'll join some friends for a drink later." If one of the women has a stronger relationship with you or is a little possessive, invite her over half an hour earlier than the other woman. Don't mention to either that anyone else will be at the house. They should assume they're coming over to sleep with you one-on-one.
6. Call each woman as she is en route to your house. Tell her, "I'm inviting a little plaything over here for us. I think you'll enjoy her."
7. Chances are that the woman won't say yes. But she most likely won't say no either. Rather than waiting for a yes answer, as soon as the idea has registered with her and she seems to have assented nonexplicitly (often by saying something cute and noncommittal), change the subject and talk about something else. Here's an example of one such dialogue with a more skeptical woman:

YOU: I've invited a little plaything over for us tonight.
HER: What do you mean? A toy or a person?
YOU: A person. Her name's [*name*].
HER: [*long pause*]
YOU: Remember, we keep talking about wanting to do something new and adventurous, but we never end up doing it. So I thought it was time we did.
HER: A friend of mine told me that bringing a third person in can really ruin a relationship.
YOU: Well, there's no obligation to do anything with this other person. It doesn't really matter what happens. We can all hang out and, if it doesn't seem right, we don't need to do anything with her. It's only if we're both comfortable and it seems right. Otherwise, we can just have her leave.

HER: I'm not sure I'm comfortable with another woman in bed with us.

ME: That makes sense, because how these things go all depends on how the guy behaves. If he makes it all about him, no one has fun. It's lame and uncool. But if he makes it about the girl's pleasure and her experience, then it can be a great, comfortable thing.

HER: [*silence*]

YOU: [*changing the subject*] Oh my God, I'm sitting outside right now and it's so beautiful. It's like the perfect temperature—not hot, not cold. How far away are you?

HER: About fifteen minutes.

YOU: Great. See you soon.

8. When both women come over, instantly engage them in an entertaining, nonsexual activity, such as playing a female-friendly video game, doing one of your value demonstrations, or helping you go through your closet to find clothes to throw out.
9. Offer them a drink—not to get them drunk or even tipsy, but just to create a romantic, transgressive mood.
10. At some point in the first fifteen minutes, excuse yourself to go to the bathroom or make a phone call. It's important to leave the women alone for a few minutes to get acquainted.
11. Shortly after you return, bring them to your bed for a nonsexual activity (such as showing them pictures or videos on your computer). Rather than positioning yourself between the women, let the woman who is more likely to be possessive (Woman #1) be in the middle.
12. At this point, the women will know what's going to happen, and will most likely begin to initiate it by being playful or suggestive. Casually start making out with Woman #1. As you do so, hold the hand of Woman #2.
13. Now make out with Woman #2. As you do, guide her head so that you are both making out just above Woman #1.
14. Now move your head away and, if necessary, gently turn Woman #2's face toward Woman #1's. Most often, they'll start making out passionately, thus beginning the threesome. Throughout, make sure that neither woman gets jealous of the attention you're giv-

ing the other—or the attention the other is giving you—even if it means sacrificing your own pleasure. One time in twenty, they may not want to kiss, but they will both want to be with you. So simply take turns progressing with each, making sure they're both always receiving at least some sort of attention from you, even if it's just eye contact.

EPILOGUE

A Note to Female Readers

You may be reading this and thinking, "Oh my God, this one guy totally did these routines on me."

You may then think, "I was tricked."

This note is to reassure you, in a non-sarcastic, non-glib way, that these routines exist only to help men avoid awkwardness and rejection.

You cannot be tricked into sleeping with someone you don't want to. On the other hand, you can very easily be dissuaded from sleeping with someone you *do* want to. These routines were designed to prevent men from scaring away or boring to tears someone they like or love or desire.

If any of this material disturbs you, just remember that the rules of the game weren't created by men. We'd love to walk up to you and say, "Hi, let's exchange phone numbers" or "let's get coffee" or "let's get married" or "let's fuck in that alley right there." But if that worked with any regularity, you'd have hundreds of guys approaching you and saying the same thing every day. So, consequently, you've developed a screening process to separate the desirables from the undesirables.

Though the routines are designed to follow your rules—not the ones in your conscious mind, but the ones in your subconscious—in the end, you make the decision. You say yes or no, stop or go. And whether or not a guy is using these routines has absolutely nothing to do with whether he's being sincere or phony or whether he's a good or bad person.

If a man is using these scripts, all it means is that he's read this book and he doesn't want to lose you due to his own nervousness or inexperience or anxiety.

So is this material manipulative? Of course it is. Every great romantic comedy begins with some sort of manipulation, even if it's just a woman purposefully dropping something in front of the guy she wants to meet or a man pretending to be more successful than he actually is. As human beings, it's our nature to manipulate. Even a baby crying is trying to manipulate his parents for food or attention. The real question you should be asking when meeting a man isn't, "Is he trying to manipulate me?" but "Is he trying to manipulate me with good intentions or bad ones?"

And if his intentions are good, you know what to do.

THE STYLE DIARIES

I here present you, courteous reader, with the record of a remarkable period in my life, according to my application of it. I trust that it will prove not merely an interesting record, but, in a considerable degree, useful and instructive. In *that* hope it is, that I have drawn it up: and *that* must be my apology for breaking through that delicate and honorable reserve, which, for the most part, restrains us from the public exposure of our own errors and infirmities. Nothing, indeed, is more revolting to English feelings, than the spectacle of a human being obtruding on our notice his moral ulcers or scars.

—THOMAS DE QUINCEY,

CONFESSIONS OF AN ENGLISH OPIUM-EATER, 1821

THE RULES OF THE GAME GOVERN OUR LIVES,
OUR PROSPERITY, AND OUR HAPPINESS.

THE RULES OF THE GAME ARE EMOTIONAL
AND NOT LOGICAL.

THE RULES OF THE GAME HAVE BEEN THE SAME
THROUGHOUT HUMAN HISTORY, REGARDLESS OF RACE,
CULTURE, OR NATIONALITY.

THE RULES OF THE GAME ARE IMMUTABLE.

THE RULES OF THE GAME CAN GET YOU LAID,
LOVED, MARRIED, IMMORTALIZED.

THEY CAN ALSO GET YOU BETRAYED, DUMPED,
DEPRESSED, STALKED, BEATEN, STABBED, SHOT.

HANDLE THEM WITH CARE—FOR THESE PAGES ARE
INTENDED NOT AS PRESCRIPTION
BUT RATHER AS PREVENTION.

CONTENTS

INTRODUCTION

"What are your goals?" he asked.

"My goals?"

"Yeah. Unless you know where you're going, you won't know how to get there."

"I guess my goal is quantity, quality, and variety. My goal is to make out with women I just met, get blow jobs in club bathrooms, sleep with a different person every other night, and find myself in strange sexual adventures with multiple women."

He sat in silence, listening, so I continued. I'd never articulated it before, either out loud or to myself. This was several years ago, just after I had discovered the Rosetta Stone of attraction in the form of an underground society of master pickup artists. "I want to corrupt young virgins, reawaken passions in bored housewives, seduce and be seduced by stars, students, centerfolds, businesswomen, and Tantric goddesses. And then, from amongst these women, I will choose one to love."

"How will you know when you've found her?" he asked.

"I guess I'll just know, because I won't want to be with other women anymore."

"Well, that sounds like a good plan. And it makes sense to a point." I waited. I knew he was about to find the flaw in my logic. "But what happens after a year or two years, and the sex isn't as exciting anymore? What happens if you have a child with her, and she becomes less available for you emotionally and sexually? What happens if you go through a rough patch and start fighting all the time?"

"If those things happened, I'd probably want to sleep with other women." I watched him as he lifted his legs off the floor and crossed them on the couch in a position of spiritual superiority. "But I'd just have to control myself. I suppose I could think of other women like cigarettes. Even though I desired them, I would refrain from indulging because I'd know it was bad for the health of the relationship."

And then I waited for it, the inevitable question. He was a music producer, yet he never seemed to work. Instead, I'd meet him at his house in Malibu, and we'd spend hours discussing the meaning of life while his Indian houseboy brought us bottles of water and plates of vegan food.

"So," he said, "you'd be okay spending the next fifty years sleeping with only one woman?"

He had walked me into the weakness in my romantic strategy, and probably in most men's. I love women's laughter. I love their lips, their hips, their skin, their touch, the way their faces look when they're in the throes of sexual ecstasy. I love the way they nurture, feel, care, intuit, understand unconditionally. I yearn to create that bubble of passion, which draws us into the moment and connects us to the energy of the universe. And I cherish, more than anything, the moment in bed right after the first time, when all that there is to hold on to has been given. "Well, that would be difficult for me," I admitted. "Ideally, I'd like to be able to have my cake and eat it."

"I think that's a reasonable request," he said. "After all, cake was meant to be eaten. Who actually orders a cake, then doesn't touch it?"

"So what you're saying is that there's a way to be in a committed, loving relationship, yet still sleep with other women?"

"I didn't say that. All I said is that there's a way to have a cake and eat it."

"How? Even a monogamous relationship is a challenge. That's why twenty-five percent of all crimes are domestic violence, that's why the divorce rate is fifty percent, that's why the majority of men and women have cheated. Maybe the relationship paradigm that's been forced on us by society isn't natural." He looked at me disapprovingly. I continued anyway. "Even if you're faithful for those fifty years, you still may check out a woman walking by or leaf through a copy of *Maxim* or look for porn on the Internet one night. And this is going to make your partner feel like she's not enough for you."

"This is true. You can't have a healthy relationship if your partner doesn't feel secure."

"Exactly. So, considering the nature of men, how is it possible to make a woman feel secure in a relationship?"

"Probably by not wanting to have your cake and eat it," he said.

"But that's not natural. You just said that cake was meant to be eaten."

"Well, then," he said, "you'll have to find a way to eat it without hurting someone you love."

I hated him sometimes. For being right.

In the days that followed, I sifted through the conversation in my mind, searching for answers. I talked to men and women everywhere I went, asking each the same question: "If you didn't have to worry about having children and you didn't need someone to take care of you when you were older, would you still get married?"

Most men said no. Most women said yes. And that was when I realized that the traditional relationship model is defined by a woman's needs, not a man's.

Then I started asking a new question:

"Let's say you met someone, clicked on every level, and wanted to date this person. But the person said that after two years, he or she would disappear from your life forever and there was nothing you could do about it. Would you still date this person?"

Most women said no. Most men said yes—some even said the scenario would be ideal.

So where does that leave the "one woman, one man, happily ever after" myth that is the basis of our entire civilization? Apparently, on an unbalanced scale, because the natural instincts of men seem to be to alternate between periods of love relationships and periods of hedonistic bachelorhood, with some traumatized kids thrown in as an evolutionary imperative.

When I next met my friend, I shared my conclusion. "That's kind of a sad way to live one's life," he said.

"Yeah, and the problem is that's exactly how I've been living mine. Except for the kids part. I don't want to traumatize them, so I'm waiting until I figure out a solution to this whole relationship dilemma that satisfies the needs of both sexes."

"You'd make a good politician," he said, not as a compliment. "You're the type of guy who can't kill a fly, a bee, or a cockroach himself, but has no problem hiring an exterminator to kill a whole swarm of them."

"What's that supposed to mean?"

"It means," he said, setting down his bottle of water, "that your ethics are fucked up."

We live in a society that likes to make clear-cut judgments—between good and bad, right and wrong, successful and unsuccessful. But that is not how the universe works. The universe does not judge. Since the dawn of time, it has operated on just two principles: the creative and the destructive. We have come to terms with the creative impulse—that, after all, is why we're here—but we live in fear of the destructive because that, one day, will be our reason for going.

I don't want to just offer you a self-help book and tell you that, if you follow it, in thirty days your life will be perfect. There's another side to the game: the destructive side. And, the more successful you are, the more you're going to rub against it. Especially since, more than any other instinct we have, the sexual impulse contains both the creative and the destructive.

The inspiration for this book was the preceding series of conversations, which point to a seemingly irreconcilable disparity between the sexual and emotional needs of men and women—not to mention a reluctance to admit and express them. They also underscore a similarity that transcends gender: the fear of being alone—and the dramas and comedies that occur because, as the director Rainer Werner Fassbinder put it, "we were born to need each other, but we haven't learned how to live with each other."

The eleven stories that follow are true, and all except two happened during the period in which I immersed myself in the pickup artist subculture and was given the alias Style, as chronicled in *The Game*. Unlike *The Game*, however, these stories are less about getting the girl and more about the nature of desire itself. They loosely trace the metaphorical arc of a man's dating life, building toward the question that none of the pickup gurus I met while learning the game was able to answer: What do you do after the orgasm?

Fiction writers are lucky: They can hide behind the flawed characters they create. Here, the only flawed character is me. In the process of approaching thousands of people to master the game and myself, the three engines driving my behavior—hereditary instincts, family upbringing, and social forces—came into constant conflict. As a result, I hurt people's feelings, made bad choices, took unhealthy risks, missed important opportunities, and committed irreversible blunders.

I also had some amazing sex.

And therein lies the conflict.

From each of these experiences, I've tried to extract a lesson. And that hasn't been easy. Because some of these experiences never should have happened in the first place.

RULE 1
ATTRACTION IS NOT A CHOICE

I am sitting on her couch and she is waiting for an answer.

She is offering me French lessons.

She is sitting too close. She is talking too slow. She is accidentally on purpose grazing my knee with the back of her hand.

She wants me.

She has to be at least sixty.

And, somehow, I feel myself drawn in.

I know the symptoms: dizzy, light-headed, eyes defocusing, room melting, PC muscle contracting.

I look at her: she is old, man. And not a good old. Just plain old. And worn-down. Brittle black-gray hair piled sloppily atop her head. Pea-size pores freckling her face. Body like a bag of gravel. Blood-pressure socks. Varicose veins. Granny glasses. Mustache.

I have to get out of here. Before it's too late.

"Gotta get back to writing . . . me, too . . . well, bye then . . .sure, a French lesson would be . . . I'm not sure when . . . work and all . . . but, yeah, definitely . . . and give my best to Josh . . . thanks . . . you, too."

Jesus. That was close.

We have lived on the same floor of the same apartment building in Pasadena for six months. We've passed each other in the hallway many times. She's always with her autistic son, Josh. I feel bad for her. She's a single mother, and has sacrificed her entire life to take care of her son and nurture his autistic musical genius. He knows the name, lyrics, chords, recording date, and catalog number of every Beatles song and is not too shy to recite them to strangers. He never forgets a face or a fact. He has aged her prematurely.

Yet every time I run into her in the hallway or the elevator, there is this tingle. This energy. I feel drawn in and hypnotized. I can't describe it any better. But I know it's attraction. I want to kiss her. It makes no logical sense. The only older women I've slept with were ones any red-blooded boy would go for: long legs, workout bodies, spray tan, shampoo-commercial hair. I've never been drawn to a woman like this before. Yet, sometimes, at night, as I prepare to sleep, my hand will lazily drift into my boxer shorts. And I'll find myself thinking of her.

I live in Los Angeles. I see some of the most gorgeous women in the world on a daily basis. They're everywhere: carrying their crappy little show dogs, sitting in Starbucks on a Tuesday afternoon because they're too pretty to work a day job, jogging along the beach like they're auditioning for *America's Next Top Model*.

And what do I do? I masturbate to the sixty-year-old crone in my building.

I could have anyone in my fantasies. And by this point, I could have just about anyone in real life, too. Why do I keep choosing her?

Two days later, I'm taking the elevator to the garage with the previous night's companion, Darcy. She is sexy but shady. Claims her job is throwing parties for men in Las Vegas. I would like to go to one of those parties sometime.

"Hi, Neil," a loud, nasal voice greets us when we step out of the elevator.

It's Josh. He met Darcy in the building once before, about three weeks ago. He just turned fifteen. He's starting to get acne and feelings around girls he can't explain. He likes to talk to me about masturbation and how he hates his mom.

"Hi, Darcy. You're twenty-six and from Newton in Massachusetts, right?" He knows he's right. Show-off. "You're pretty."

Nancy weak-smiles at us. "I'm sorry. Josh, come on."

I look at Darcy. She is tan from a bottle. She is buxom from a Beverly Hills doctor. She is rail-thin from crystal meth. She is a porcelain doll of youth, sexuality, and doom.

I look at Nancy. She is pasty from indoor lighting. She is saggy from age. She is lumpy from lack of exercise. She has given up on youth, on sexuality, on herself. The autistic cross she's had to bear for so many years has consumed her, broken her, wrecked her.

What was I thinking?

"Hey, Neil, 'The Long and Winding Road' is a good song. Do you like that song?"

"It's great," I tell Josh.

"It was written on the same day as 'Let It Be,' " he informs me. "It's the only song on the album that just has Paul McCartney on piano and not Billy Preston. What do you think he means when he says, 'crying for the day'? What day is he crying for?"

That's the tragedy of Josh. He knows facts. But metaphor is too vague.

"The day when things were better."

"Don't you think he could just mean the day before?"

He is too literal. He doesn't realize that if words only represented their dictionary definitions, they would no longer serve the purpose of expression. There would be no Beatles, no literature, no poetry. There is something underneath each word that affects its expression and interpretation. That thing is called emotion. The inability to recognize it is something both Josh and Darcy have in common.

"Josh, let Neil go," Nancy coos from inside the elevator, finger mashed against the open-door button. Then to me: "He's excited because he's going to stay with his piano teacher tonight."

The door closes. And I wonder what she meant.

Was she just apologizing for his behavior?

Or was she trying to let me know she's going to be alone tonight?

I can't even say for sure that she's ever thought about me in that way. And, surely, after seeing Darcy, she can't expect I'd actually be interested in her.

The whole thing is just ridiculous. I see a lot of potential in Josh, though; I'd like to turn him on to more good music. It would be nice for him to have a mentor closer to his age.

That night, I find myself on Nancy's doorstep. There is a Zombies CD in my hands. I keep telling myself I'm just dropping off a CD for Josh, because I think it'll open up a new world of music for him. But I know why I'm really there: to see what happens.

I don't think I would actually go for it, if given the chance. That would be gross. I just want to satisfy my curiosity. And she seems interesting as a person. Very cultured. I'd like to know about her background: What she was like before she had Josh. How she makes a living. Where she learned French. Stuff like that.

Nancy doesn't seem surprised when she answers the door. She is wearing a black shapeless dress and lumpy stockings, and her cheeks are awkwardly rouged. The sleeves of the dress cinch her arms above her elbow, creating a roll of skin that reminds me of a Polish sausage.

She steps to one side and holds the door open. According to the rules of politeness, I must enter.

Now I am in the lair. And I feel the energy shift around me.

"This is for Josh," I tell her.

She takes the CD from me. Her fingers don't touch mine.

"Would you like some tea?" she offers. "I just made some."

The web is forming.

"Sure."

I sit on the couch. It is burlap, with a yellow-and-white–knit blanket thrown over it. It smells like sandalwood and ashes. I'm having trouble breathing. My chest is tightening. I look at the door. It seems so far away.

I am sunk.

My dick is pushing lightly against the denim of my jeans. What is going on?

I look at Nancy. My grandmother was a prettier lady than her. This doesn't make sense.

She shuffles over with the cup of tea. I thank her.

"*Je vous en prie,*" she responds.

I love it when she speaks French. Her accent is perfect.

We talk about Josh. That is all we ever talk about. He is practicing for a piano recital. He can figure out any song by ear. His teacher is impressed. I can't focus. I can't focus. I can't focus.

She wants to show me pictures. They are in a cream-colored album. She sits next to me, lays it in her lap, and opens it with elegant fingers. The front cover drops onto my left leg.

"This is Josh and his teacher standing outside the Schoenberg Music Building."

I don't see. I don't know. I don't care. My nostrils are filling with her scent. My heart is hammering. The room is spinning. I need to do something to stop it.

I raise my hand and clumsily brush a stray strand of hair off her face. It feels like a pipe cleaner.

She stops speaking, lifts her head, and turns toward me. A blast of sandalwood ash hits me in the face. I must have her.

My lips crush hers. It is like the triumphant last chord of a symphony ringing in my head.

Her lips are rough and bumpy, but her tongue is soft and fat. She just sort of puts it in my mouth. It lays there, and it feels nice. It emits that slow, sensual energy she has, sending it all through my body.

I know this is wrong. I'm fully aware that a line has been crossed.

Fortunately, she senses that I'm uncomfortable.

"Should we go into the bedroom?" she asks.

I am not shocked by this. I actually think that it is a great idea.

She leads the way. I follow, and as I see her body moving in front of me, bulging everywhere with no shape that could be defined as sexual, the spell breaks. For a moment, I have the option to leave. But I don't.

I am compelled by my own nature to finish what I start. And perhaps I never really had the option to leave anyway.

She sits on the edge of what looks like a hospital bed. With effort, she slowly raises her legs off the floor and onto the mattress.

I remove my shoes and join her. She doesn't say anything and neither do I. One word would ruin it.

Her hands wrap around my back. Our tongues reunite. The smell of old lady oozes from her skin. I do not want to take this slowly.

I start to pull off her dress while balancing on top of her, then roll off and let her finish the job.

Her skin is the color of oxidized newspaper. Her underpants end where her bra stops. Both pieces seem excessively large. And they do not match. The underwear is white, the bra is what they call nude. They are about function, not form.

I do not want to linger here. I do not want to linger anywhere.

I reach behind her and release the bra hook by hook. I place a breast in my mouth. It seems like the right thing to do.

I am able to disconnect for a moment, to imagine her as desirable as I circle my tongue around her nipple. Encouraged, I decide to stop looking and retreat into the world of feeling.

But then I reach down to slide off her underwear. And beneath, instead of skin, I feel plastic. I grope around it. There is some sort of plastic bag attached to her side.

I can't remember much after this. I recall a strange wave of nausea coming over me. I recall proceeding anyway because it is my nature. I recall it lasting no more than five minutes. I recall making the minimum amount of necessary conversation afterward to ensure her comfort. And then leaving.

In the days that followed, I didn't think about Nancy much. Not in the way I used to. I talked to her on the phone a couple times afterward, just so she wouldn't think I was avoiding her in the hallway, which I was. I can't say why I no longer fantasized about her. Maybe it was that I'd attributed a certain sensuality to her that didn't exist in reality. Or maybe it was the plastic bag.

A month later, I moved out of the building. Not because of Nancy. Because I felt isolated and listless in Pasadena. I wanted to live where people were struggling and striving and trying to become, because that's always where the action is. That's where you find life. That's where you find beautiful, desperate women, if that's your sort of thing.

I called Nancy and said good-bye. I promised to stay in touch and see Josh's upcoming recital.

That's where the story should end. But it doesn't. In fact, it probably shouldn't have even begun. Nonetheless, seven months later when I was collecting my mail from the building manager, I saw her again.

She looked thin. She'd lost at least thirty pounds. Her hair was clean, dyed black, and tied in a perfect bun atop her head. She was wearing lipstick, mascara, eye shadow. She practically glowed.

On her arm was a man. He appeared to be her age. Small and bald, but not bad-looking. He was sprightly, well-tanned,confident.

"Hey, you look great," I told her.

"Merci." She seemed happy.

"Where's Josh?"

"I moved him to a different floor," she said in that slow voice that had once charmed me. "He lives in apartment 502 now, with a tutor I found for him."

She fell silent for a moment and smiled thinly at me. She'd even bleached the hair over her lip. *"Merci,"* she repeated.

There was a new energy around Nancy. It wasn't attraction. It was gratitude. I felt like I'd done something nice for her, that I'd unlocked and released something she'd forgotten she had. Perhaps that was the energy I had felt the whole time: an exuberant woman trying to break free from the prison she'd been in since her son was born.

I thought for a moment that maybe I'd found a calling: the angel of fuck. There are, everywhere, women who have given up on their sexuality. I see them in the airport, too scared to break up with the cheating husbands who take them for granted. I see them at the beach, so busy tending to their ungrateful children that they've forgotten to tend to themselves. I see them at the twenty-four-hour diner, still nursing the wounds of a breakup that happened decades ago, watching the twenty-year-old waitresses with hateful eyes, thinking, "Someday. You'll see."

They were all once eighteen and bursting with youth, spirit, sensuality, possibility, and countless potential suitors, one or two or ten or twenty of whom would drain away all that light. I could seduce them. I could slowly, tenderly fuck each and every one of them. I could make them eighteen again. Not for me, but for them. So their sexuality, their passion, their selves could reawaken, and they'd realize that life still lay ahead of them and eighteen wasn't all that great a year anyway.

I could do that.

As I left the house, climbed into the secondhand SUV I'd just bought, and drove back to my new place in Hollywood, I realized the flaw in my plan: it wasn't me who had seduced and saved Nancy. She had seduced me. And I'd moved. I'd changed. I'd grown up.

Maybe the gratitude I felt was my own.

RULE 2
ONE BROKEN LINK DESTROYS THE CHAIN

Kevin is going to be here any minute. He wants to go out and meet women. And I'm still in my boxer shorts. I have not showered or shaved in days, man. When I look in the mirror, I see the ghost of Yasser Arafat staring back at me.

I should not be going out when I have a book due in fourteen days. But my eyes are going to melt in their sockets if I keep staring at this computer. I've been writing for three weeks straight. It's time to interact with living beings again. My social skills are rusting.

Have to get my act together quickly. My lucky broken vintage Vostok Soviet military watch has somehow time traveled into the kitchen, where it's lying facedown in peanut butter. I need to clean the kitchen. It could be embarrassing if anyone came back here.

I should add that to my list. But first I need to find the list, which is probably in the pocket of my Levi's premium boot-cut jeans. The jeans are in the clothing pile. This is where items go that I've worn but don't smell bad enough to get cleaned yet. It is an altar from which I compose my identity every day.

I had an idea last night for a book that I also need to add to the list. What was it? Something about living without technology for a year.

Shit. There's the buzzer. It's Kevin. Forgot he was coming and he's already here. Get it together, Neil. Kevin needs you to be his sacrificial lamb and start conversations with the beautiful women of Southern California.

Grab Levi's premium boot-cut jeans. Smell jeans. The scent is a cross between macadamia nuts and my room after sex. That'll work.

"Hey." Kevin grins lopsided when I answer the door. "You going out like that?"

Putting on other leg of jeans now. Just have to find a shirt. Something cool. Something from my pile, because if it's cool, I've probably worn it in the last month. And if I've worn it in the last month, I definitely haven't washed it.

Fish for black shirt. When in doubt, wear black. It's the safety net of male fashion. Grab tail of gray knit tie I bought in London and pull loose from pile. The tie looks puffy. I may have accidentally washed it last month.

Just need a belt. Must sort through pile to find belt. Every item tells a story. This yellowy T-shirt I picked up seven years ago at a Boston warehouse that sells clothing for a dollar a pound.

"Hey, man, it's gonna be crowded if we don't get there soon," Kevin says. He shows up late and he's mad at me, like I'm some kind of dawdler.

Just use puffy gray tie as belt. Now need something around neck. Pendant necklace? Too disco. Shoelace? Too thin. Red ribbon from Christmas present? Fine. It's like nature's own silk tie.

"Ready?"

"Ready."

"Like that?"

"I'll be fine. I can rely on my charm."

Kevin is my friend, but not really. If my car broke down, he's not someone I would call to fetch me. We are united only by our shared pursuit of women.

"Remember the girl I had call you the other night?" he asks as I unlock my car door. Somewhere underneath these Coke bottles and Red Bull cans, there is a driver's seat. "I brought her home and we were gonna get in the Jacuzzi, but my mom fucking drained it." There's precious, life-giving Red Bull left in this can. Need my taurine. "So we got in anyway, and she gave me head while I looked at the stars." Kevin is sitting on my rough draft.

Feel like there's fog in my head. Gotta clear it out. Get present in the moment. Clap my hands. Shake my head. Use my voice box.

"Testing, testing." It works.

"What are you doing?" Kevin asks.

"Warming up."

Drive 2.3 miles to James Beach bar, hand valet keys, smile, enter, pretend to be normal. Girls everywhere, drinking, laughing, each one unique and growing ever more intoxicated by the sudden smell of macadamia nuts in the room.

Two women who appear to be in their twenties walk away from the bar. Must start talking or I'll be stuck in my head all night. I feel Kevin's hand on my back pushing me toward them. I should package Kevin's hand and sell it to men who are too scared to approach women.

"Have you met my friend Kevin?" I ask. "He's in the world's only all-Jewish Christian rock band."

"A what?" asks one of the girls. Model tall, stringy blonde hair, sand-dollar complexion, white jacket with rainbow buttons. Seems like the kind of girl you'd meet at one of those bookstores that sell incense at the cash register.

"He's in a band," I repeat.

"So am I," she says. She is friendly and kind of sweet. I didn't expect her to take me seriously. I suppose rainbow buttons are a sign of tolerance.

Her friend has a tight white tube top, compact frame, long black hair, angular face. The kind of girl you'd meet in the sales office of a gym.

I need to start going to the gym again. And eating healthier. And flossing every night. I'm losing it all.

"Is that peanut butter on your watch?" Bookgirl asks, touching my hand.

"Don't manhandle it. It's vintage Soviet military peanut butter. Worth a fortune."

As Kevin and I talk to Bookgirl and Gymgirl, we automatically pair off. Why do I bother to write? This is so much more fun.

"You have one life to live." I hear myself telling Bookgirl. The words are not mine. They belong to Joseph Campbell, dead professor of mythology. "Marx teaches us to blame society for our frailties, Freud teaches us to blame our parents, and astrology teaches us to blame the universe." The fog has lifted. It's funny how quickly it comes back. I constantly forget that people tend to be polite, unless they think you want something from them, which, of course, we do. "But the only place to look for blame is if you didn't have the guts to bring out your full self, if you didn't act on your desires, if you didn't take advantage of what was in front of you and live the life that was your potential."

There are tears in her eyes. Thank you, Joseph Campbell. I take her hand in mine and she squeezes it warmly. Forgot to clip my nails. Have to add that to the list. I keep a list in my head of things I need to add to the list in my pocket.

"That's just what I needed to hear," she says, and takes another sip of beer, "because I'm three months pregnant, and I'm just asking a lot of questions right now."

For some reason, I am not fazed by this. I look at Gymgirl. Kevin is massaging her shoulders and whispering in her ear. I make out the words "anal sex."

Bookgirl tells me she lives with her boyfriend and loves him very much. She tells me her friend is married and has two children and loves them very much.

The night is dark.

I was introduced to Prince once in a bar, and he asked me what I did. I told him I wrote books. He asked what they were about, and I said they were about the dark side. "Why the dark side?" he asked.

"Because it's more interesting," I told him.

"But the light side can be interesting, too," he admonished.

I wish Prince were here right now. He would see that he was wrong. Every adventure to be had in this room is on the dark side. The people on the light side are asleep right now. And they are dreaming about the dark side. Because the more you try to repress the dark side, the stronger it gets, until it finds its own way to the surface. I sleep well. I dream of angels and sponge cakes and panda bears. I don't see the dark side until I open my eyes. And, tonight, it seems the dark side is going to be a pregnant New Age Amazonian who lives with her loving boyfriend.

"Will you take us to our car?" Gymgirl asks when the bar closes. "We don't like walking alone late at night."

"That will cost extra," Kevin tells them. They don't laugh. "Just wait a sec while we find our friends."

Of course, we have no friends here. This is Kevin's way of getting me alone to make a plan. And that is great. Because I enjoy plans.

"Okay," I conspire with him. "Let's tell them that our friends left without us, and we need a ride home."

"Love it. What about your car?"

"We'll just leave it with the valet and pick it up tomorrow."

The girls agree to take us home without hesitation. A simple plan can make all the difference between going home with company and going home alone.

We're walking down the street now, arm in arm. We are saving them from criminals. They are saving us from taxicab drivers. It's a fair trade.

"Wow, it's funny how we paired off into couples," Bookgirl says. My head reaches her shoulders. And if she doesn't care, I don't care.

Their car is a BMW convertible, which indicates that they surely could have afforded the valet. Maybe they also had a plan.

Bookgirl wants to play me her music. This concerns me, but it also allows me to proceed with stage two of our plan.

"This sounds great," I tell her. It is sappy and makes me want to punch butterflies. "But it's too windy to hear your lyrics. Just bring it upstairs and we can play it where it's more quiet."

She agrees.

Women are not stupid: She knows what she's just agreed to. We park and walk arm-in-arm to my front door. Infidelity is in the air. It is dark and smells like macadamia nuts.

I reach into my pocket to grab the keys.

They are not there.

I double-check my pockets, as if everything's just fine. Give myself a full-body pat down. I feel the potential of the evening begin to dissipate.

The girls are looking at me suspiciously now. All the doubts that liquor and smooth talk held back are creeping to the surface of their minds with each passing second. They know something is up.

Okay. No need to panic. Obviously, I must have my keys because I drove to the club. Otherwise . . .

Fuck. I'm an idiot. I valeted the car. So the valet still has my keys. And I'm locked out.

In the blink of an eye, I develop a plan. There's always a plan.

"I left my keys upstairs," I tell the girls. "But it's no problem. I'm just going to climb up to the balcony. I always do this."

I never do this.

"What floor do you live on?" Gymgirl asks. Good question.

"The third. Just wait right there. I'll be back in a second."

I run to the side of the building and look up. This is possible. It's just a puzzle. And every puzzle has a solution.

Gotta think quickly. I'm losing them.

I believe I can make it. No problem. If I fall, I die.

The girls follow me and look up the side of the building doubtfully. "I'm getting kind of tired," Bookgirl says. "I should probably go home."

I suppose this makes sense. After all, she is pregnant. And I really should not be having sex with her.

"This'll only take a second," I tell her. "Just wait at the front door, and I'll be right there to let you in. Don't worry about it."

It is time to save the night.

I climb onto the first-floor railing. It's loose and shakes beneath my feet. I did not plan on this. Have to move fast.

Grab the bottom of the second-floor balcony and pull myself up. Forearms shaking. Shouldn't have stopped going to the gym. Kick my legs over. A little winded. Take a short break here with the rear of my Levi's premium boot-cut jeans hanging in the air.

Okay, just have to pull my upper body up now. Quietly. If I wake anyone, they may call the police. Or shoot me.

On the second floor now. Everything is under control. Just repeat, and I'll be on my balcony and home, having sex with this girl and her embryo.

I stretch and grab the base of my balcony railing, then hoist myself up and kick my legs onto the ledge. I am almost home. Just need to pull my body up so my jeans aren't hanging in the air.

There is a slight problem. I can't move. My tie-belt is caught on something. Can't see it from this position. Probably a nail.

Must use brute force. I pull hard on the balcony railing. Forearms getting tired. Now the railing is bending toward me. This is not good.

They really make strong ties in London.

Think, Neil. Think. You're smarter than this nail.

There is a hotel across the street. Maybe I can signal to someone in the window. But what would they do? Probably just call the fire department and make a big scene.

Need to retrace my steps. Unclimb the building.

I lower myself back to the second floor and the tie slips off a rusty nail that probably once held a planter.

Standing on the second-floor balcony, I remove the tie-belt and stuff it in my pocket. The jeans slip halfway down my ass. Won't be able to climb with pants falling off. Need to remove them.

I take off my boots, step out of my Levi's premium boot-cut jeans, lean over the edge of the railing, and toss them up to my balcony.

They plummet to the pavement below.

When I look down to see if the jeans survived, I notice headlights in the street. It's a convertible. The girls are leaving. The night is ruined. I knew I should have stayed in and written. Why do I let Kevin talk me into these things?

"It's okay," Kevin yells, as I'm putting my boots back on. "The married girl is coming back."

He is talking way too loud. He's going to wake the whole neighborhood.

"I think we can double-team her," he shouts.

"Shh," I admonish him.

A light inside the apartment I'm standing outside flips on. And I'm on their balcony in boxer shorts and one boot.

There is only one way to save the situation. I race to the railing, climb on top of it, then spring onto my balcony. It all happens so fast, and in such a panic, that I don't even know how I did it. I may have just proven the theory of evolution. Surely, if I can access the climbing genes of my ancient monkey ancestors, I can live without technology for that book idea.

What a horrible night. And my room is a mess. Clothes are everywhere. My heart is hammering. Gotta remember to get my boot off the downstairs balcony later.

And pick up my jeans from the street.

And retrieve my keys and car from 2.3 miles away.

Have to add all this to my list. But first I absolutely must check my e-mail. Something important could have arrived that I may need to deal with. The glow of the computer screen and grinding of the hard drive soothes my nerves. This is where I belong. It's a jungle out there.

Kristen is coming to town and wants to stay with me. Magnus wants me to meet some Norwegian rappers. And Stephen Lynch wants me to send clips of an article I wrote about him.

I have a book due in two weeks. I can't possibly do any of these things. So I write and tell Kristen I'm working on a book, but she can stay as long as she understands that I need to write. I tell Magnus that I'm working on a book, but I can meet them really quickly for dinner, since I need to eat anyway. And I tell Stephen Lynch that I'm too busy to send his clips right now.

Clip my nails. Must add that to my list right now before I forget again.

The buzzer. Who could that be at this hour?

"What the fuck are you doing up there?"

"I'll be right down."

Kevin is sitting in front of my building. He is not happy with me. I'm probably not the kind of friend he'd call if his car broke down.

"Take that ribbon off your neck," he snaps. "You look ridiculous."

We wait and wait and wait. Gymgirl returns, then tells us she's tired and wants to go home. And I'm okay with that. After all, she is married. And we really should not be having group sex with her.

Sometimes mistakes happen for a reason. I need to write my book anyway. It's due in fourteen days. Actually, thirteen days now.

And a book is a lot of work. It requires a massive amount of organization and planning. Fortunately, these are things I'm good at.

RULE 3
GAME IS A BORDERLESS STATE

I am writing this in case anything bad happens.

If I disappear, please come looking for me.

Just remember the name Ali Raj. He's a magician, but he may have an illegal sideline. He's supposedly friends with the prime minister's son. And on the off chance that I'm breaking some taboo here, I want you to know what happened.

I love the game. And I believe I may be an addict. It's changed my life in ways I never thought possible. In high school and college, my friends came back from winter and spring breaks talking about their vacation hookups. I never got anything on vacation but a sunburn and a refrigerator magnet. I was never able to just relax and have fun. I was too busy worrying about what everyone else thought of me.

But once I learned the game, everything changed. Wherever I went, new adventures beckoned. I visited Croatia and ended up having sex in the ocean with a nineteen-year-old who hardly spoke a word of English. I flew to a small town in the Midwest for a *New York Times* article and fooled around with a rich housewife, then slept with her niece. And on my first night in Sweden, I met a girl who stripteased to ABBA in my hotel room as foreplay.

Now I'm in Bangladesh, where there are no clubs, no alcohol, and no dating. And I have options.

But I don't know the rules here. And I'm worried that I'm about to get myself killed.

I'm staying at the Dhaka Sheraton. The only other person who knows me here is my traveling companion, Franz Harary, the illusionist. He has longish blond hair, usually wears yellow shirts with puffy patches on the chest, and has a very gentle demeanor. Think Yanni with magic tricks.

He thinks I'm sick right now.

But I'm in my hotel room, waiting for Tripti to arrive, hoping that Ali Raj and his henchman don't get here first.

Here, really quickly, is how this all started:

Harary is here at the invitation of Ali Raj to perform at the First International Magic Festival. I'm here working on a book that I haven't told anyone about. I've been traveling the world in search of people with powers that defy scientific explanation. I want to find real magic, proof of the existence of the unknown, something to believe in. And there's a village on the outskirts of Dhaka, the capital city here, populated by a small tribe with a blind elder who can supposedly perform miracles on command.

Both the festival and the village are frowned on by local authorities. Bangladesh is largely a Muslim society and, as such, considers magic and miracle working a sin. According to strict Islamic law, these acts are punishable by death. Importing magicians from all over the world is a luxury that only a man like Ali Raj, with a lot of money and high-level government connections, could have made possible.

We first saw Ali Raj himself when we cleared customs. Lean, with perfectly feathered black hair and a dark walking suit, he reminded me of a wax statue of a matador. I don't believe he ever spoke a word. Trailed by a motley entourage of magicians, goons, relatives, and cologne-splattered men who identified themselves as traders, he led us to a press conference that had been set up in an airport waiting room.

The reporters clustered around Harary, who made a Coke bottle—the symbol of America—vanish for the cameras. The reporters were amazed, but Ali Raj was not. He nodded to one of his henchmen, a fat-faced Bangladeshi with a fanny pack, who ended the press conference.

Raj's men herded Harary and me into a minivan with the magicians. As we drove through the crowded streets of Dhaka, women with missing teeth and bleeding gums, men with fist-size tumors on their faces, and children with club feet and shredded lungi skirts swarmed the van at every red light, begging for change. And though the poverty was appalling, the people in the street seemed happier than the average middle-class American. I suppose if you've never had anything, you don't have anything to lose—just surviving is an accomplishment. At home, we tend to take unlimited upward mobility for granted.

I saw Tripti for the first time in the hotel lobby as I was returning to my room from breakfast the next morning. She stood out not just because she was the only female in sight, but because she was wrapped in an immaculate all-white sari with a matching sequined shawl around her neck. She had long black hair, the full lips of a supermodel, and large, round breasts that seemed to lift the fabric away from her body.

She was standing with Ali Raj, so I assumed she must be his wife and I shouldn't be staring at her breasts.

Raj, as usual, didn't speak. "Harary?" she asked through perfectly formed lips.

"He's up in his room working on the helicopter vanish," I told her. Raj translated, and we entered the elevator together.

"I like," she said, touching my earrings.

The earrings are silver spikes I bought after learning about a concept called peacocking. The idea is that, just as the peacock spreads its colorful plumage in order to attract the female of the species, so, too, must a man stand out in order to attract the opposite sex. Though I was initially skeptical, once I began experimenting with these items, as obnoxious and uncool as they seemed, the results were immediate—even in Bangladesh.

She gestured to my shaved head and asked, "I touch?" Without waiting for an answer, she rubbed her hand warmly on my head. Women in Bangladesh rarely get this physical in public with men. Her touching my ears and head was the equivalent of a woman grabbing your crotch in an elevator in America.

I led them to Harary's room and took my leave as he gave Ali Raj his requirements for the illusion—a helicopter, a pilot, a field, and a helicopter-size sheet.

For the rest of the day, Tripti sat at a table in the hotel lobby, selling tickets to the magic show with the rest of Ali Raj's team. Every time I walked past, she shot me a lingering glance that conveyed an invitation to so much pleasure.

So I decided to accept the invitation.

"Why don't you take a break and get some lunch with me?" I suggested.

She looked at me sweetly and smiled blankly.

Translation: Keep it simple.

"Lunch?"

As she tried to explain something too complicated for broken English, a short, muscular Bangladeshi man with black hair and a red shirt arrived with two Styrofoam dishes of some rice concoction he'd bought in the street.

I introduced myself. "I am Rashid, my friend," he replied. "I am cousin to Tripti."

"Do you also work for Ali Raj?"

He nodded in the affirmative. Everyone works for Ali Raj.

I suggested that we all eat together upstairs. If I couldn't get her alone, at least I could win the trust of her cousin. This was Bangladesh: I wasn't expecting to get very far anyway.

I took them to Harary's room and sat with them on the couch. Tripti's cousin politely handed me one of the rice dishes. I tried a small spoonful, and some sort of hot, deadly venom seared my internal organs.

"You like, my friend?" he asked. It's interesting how whenever someone calls you his friend when you're not really his friend, it sounds malicious.

"It's great," I choked.

Sometimes, in the heat of passion, there's a temptation to have sex without a condom. At that moment, I felt like I had performed the culinary equivalent: every guidebook warned against eating street food in Bangladesh.

Between the sexual energy emanating from Tripti, the brutal spiciness of the rice dish, and the awkwardness of the situation, beads of sweat began sprouting on my forehead. It was ridiculous to think I could have an affair with this girl. Our cultures are too different when it comes to dating and sex. We prefer premarital sex; they prefer arranged marriages.

I decided to cut my losses and take a nap in my room. This just wasn't worth risking days of diarrhea.

As I rose to leave, however, Tripti turned and whispered something in her cousin's ear. He nodded, then she stood up to join me.

When I walked into the hallway, she followed. So I led her to my room, uncertain of what she wanted or expected.

As we entered, I was mindful to leave the door open so she didn't feel uncomfortable. I wanted to demonstrate that I understood the morals of her society.

I sat down on the bed and she maneuvered into position next to me, too close for conversation. Suddenly, diarrhea seemed like a worthwhile risk.

I've seen many Bollywood movies, and one of the strangest things about them is that the hero and heroine never actually kiss. Instead, they just come excruciatingly close to doing so all through the film. So I stroked Tripti's hair. She didn't flinch. I looked her in the eyes and brought my lips close. She smelled like muscat, like desire, like something forbidden.

Suddenly, she pulled away. Then she stood up and walked toward the door. Perhaps I'd been too forward and misinterpreted her actions.

Instead of leaving, however, she closed the door. "I like you," she said as she walked back toward the bed.

Evidently she was more a fan of Hollywood films than Bollywood—which are Indian anyway. So I threw her onto the bed and we began making out.

This was where things began to get weird. I realize they were already weird, but they got weirder.

She placed my hands on her breasts and began speaking in a stream of fractured Bengali-English. It came breathy, in my ear, difficult to make out. All I could catch were the names "Bill Clinton" and "Monica Lewinsky."

And this completely confused me, because I wasn't sure if she was offering me a blow job using the only English words she knew as a synonym, or if she was simply sharing her thoughts on American politics.

Assuming the best, I decided to try to remove her sari. Never having actually removed a sari before, I wasn't sure where to start.

She shivered with pleasure as I fumbled around her neckline, then she yanked my hand away. "I good girl," she said. "It is okay. I like you."

Translation: "I don't normally do this, but actually I do normally do this. I just don't want you to think I normally do this."

She unbuttoned my shirt and ran her fingers along my chest. Her other arm leaned directly against the bulge in my pants. Then she began whispering, over and over, sensually. At first I thought she was saying "*cholo*." But the tenth time around, I sounded it out as "*chulatay*."

Every cell in my body was vibrating with desire for her, while every cell in my brain tried to compute how and why this was happening.

Three *chulatays* later, she disentangled herself, straightened her sari, and stood up as if nothing had just happened. "No person," she said as she put a finger to her lips.

Translation: Either "Don't tell anybody" or "I will kiss no one else because we're now engaged."

Then she said the two words that struck fear in my heart, "Ali Raj," and made a slashing motion over her neck.

"Good girl," she repeated.

I knew I was in over my head. Yet something inside propelled me to proceed. Perhaps it was the same impulse that compels a child, when someone draws an imaginary line in the grass with the toe of his shoe and orders him not to cross it "or else," to gingerly dip his foot on the forbidden side of the line in response. It's not just an act of defiance, it's a call for adventure. His side of the line is boring; the other side contains the unknown, the "or else." The Ali Raj.

While waiting for the festival to begin that night, I made it my mission to find out what *chulatay* meant. I eventually narrowed it down to one of two interpretations: either "hanging" or "I'm hungry." Hopefully, the latter interpretation was correct.

That night, the streets around the magic show swarmed with police and reporters. The theater was in a university neighborhood, the center of Islamic radicalism, and there had been several bomb threats. Every time someone bicycled past with a package in his handbasket, I imagined the next day's headline: "Terrorists Make Magicians Disappear." Nonetheless, I headed inside. Who wants to live in a world without magic?

I found Tripti walking through the foyer and led her to the back row. As an illusionist from Spain named Juan Mayoral performed some sort of magical love soliloquy to a wire mannequin, Tripti took hold of my inner thigh. She squeezed it and, her breath wet in my ear, whispered, "How is Babu?" She then began rubbing Babu through my pants.

I looked around the theater: there were Bangladeshi men everywhere and a few scattered families. Everyone was staid, mannered, reserved, intent on the show, and I had this Muslim girl moaning in my ear. Every man has his secret fantasy: This, I realized, was mine.

As happens with most fantasies, however, reality soon intruded. The fanny pack-wearing Ali Raj henchman from the press conference plopped down in the seat next to me. Tripti quickly withdrew her hand.

"Are you married?" he asked. He knew exactly what was going on.

"No," I told him.

"Will you marry her?"

"I just met her." I couldn't tell if he was cockblocking, or if this was all some kind of plan to marry Tripti off to an American.

Between acts, I decided to try to find a secluded place to take Tripti. There were all kinds of stairwells and rooms backstage. But when we stood up, Fanny Pack rose with us and cleaved closely to our sides.

"My friend," a voice greeted me as I walked into the foyer with my growing entourage. It was her cousin. My enemy. All men here were my enemies.

He threw his right arm around my shoulder. "This is the American writer," he said to three nearby men, who were either family or Ali Raj henchmen or both. They circled me and all began friending me at once. Whenever I craned my head to look for Tripti, they redirected my attention to their conversation: "Is this your first time in Bangladesh?" "How do you like Bangladesh?" "You must come to my home for traditional Bengali dinner."

Finally, I caught sight of Tripti, who seemed either oblivious to her protective barrier or pretending to be in order to preserve her honor. I whisked her into the theater, but the phalanx of Bangladeshi men followed, tripping to get ahead of us, between us, alongside us.

When we sat down, they arranged themselves everywhere around us. Fanny Pack motioned for Tripti to move over, took her seat, and spread his legs until his knees touched mine. It all felt malevolent. As if, instead of fighting, they just got real friendly here.

"So you like Tripti? Maybe you meet her mother and father?"

Just then, I felt a sharp kick in my abdomen. I doubled over with pain.

The spicy rice had done its damage.

That night, I returned to the hotel in defeat. I spent the next hour on the toilet letting go of my need to get laid in Bangladesh. In the morning, I popped an Imodium so I could visit the miracle village with Harary later that day.

In the lobby, I saw Tripti in her usual spot at the ticket table, looking radiant in a heavily beaded all-black sari.

"Ali Raj say no leave table," she said fearfully.

I was dumbfounded by the degree of effort these men were making to keep us apart. It was as if we'd been swept up in some epic romance: two lovers from different cultures separated by family—and an evil magician.

These obstacles only served to intensify my desire for her. So, like a fish compelled by hunger toward the worm of its own doom, I made a desperate move and did one of the most clichéd things I can lay claim to in a long tradition of clichéd behavior in pursuit of women: I handed her the key card for my hotel room.

"Tonight, no magic," I told her. "Come here. I wait."

"But Ali Raj," she protested. I was sick of hearing those two words.

"No Ali Raj," I said. "You. Me. Tonight. Last chance."

I sounded less like I was seducing her and more like I was having a going-out-of-business sale.

After a moment of reflection, she responded slowly, gravely, "Okay, I come."

To give her a plausible excuse to visit me, I purposely left my sunglasses lying on the ticket table. It seemed romantic in a sleazy sort of way.

Then I walked out of the hotel to join Harary in the van scheduled to take us to the miracle village. The only problem was that the trip had been arranged by Ali Raj. Everything was arranged by Ali Raj. So the van was full of my new friends. The only one I felt I could trust was a sweet older magician wearing a polyester suit two sizes too large for him. His name was Iqbal.

Fanny Pack took a seat next to me, threw his bullying arm around me, and asked, with a slow smile and wink, "You sleep well, my friend?"

"Fine," I muttered. I wanted to get away from him. This friend shit was clearly the Bengali equivalent of Chinese water torture.

"What is this?" Fanny Pack asked, reaching across with his other arm to touch the zipper on my jeans.

"Dude, what is your problem?" I leapt up and took a seat next to Iqbal. Cockblocking I understood, but cocktouching was completely new to me.

"If we were in America, I'd smash his face in," I told Iqbal. Their head games were clearly getting to me.

"The men here like to control the women," he said patiently. "There are more acid attacks in Bangladesh than any other country."

Acid attacks?

"Yes, when men throw acid in the faces of women who reject them. It is better now because of strict laws."

Bangladesh had successfully beat me. Scared me away from its women. It wasn't worth risking Tripti's disfigurement just so I could have a local girlfriend I'd never see again. I was in no shape for sex anyway: My stomach felt like it was trying to digest a sea urchin shell. I needed to find her when we returned and call off tonight's escapade.

After another hour and a half of bumpy, bowel-jiggling roads, we arrived at the village, a collection of crudely painted shacks in a barren field of dirt. No one had digital satellite TV or a subscription to *InStyle*, so we were the entertainment—especially since Harary had brought a film crew to capture him fraternizing with the locals.

The women were beautifully made up and covered head to toe in jewelry. As we walked around, I noticed a group of teenage girls following me and staring. Eventually, a few worked up the courage to approach and began gesturing to my earrings, bracelets, rings, and shaved head.

I asked Iqbal to talk to the women and find out what they were up to. "All the women, they like you," he came back and told me. Then he pointed out a pair of barefoot, bejeweled beauty queens and said, "Those girls want to marry you."

"Why don't they want to marry Harary? He's the one all the cameras are following."

Iqbal talked to them a moment, then turned and smiled. "They like you."

In that moment, I learned that the game is universal. Peacocking—the rule of standing out rather than fitting in, of embodying a more exciting lifestyle instead of the one people are used to—seems to work in every culture. I was now officially doomed to dress ridiculously for the rest of my single life.

When we met the miracle-working village elder, I discovered something else that was universal: the principles of magic. Her miracles were just sleight-of-hand tricks, originally and masterfully executed using chicken bones. We then watched a snake charmer antagonizing a snake that had been devenomed, and a man performing an old fakir trick in which he swallowed a string and then appeared to pull it out of his stomach.

So what we discovered was not people with powers we couldn't explain, but a village of magicians who've passed down tricks from generation to generation—and who travel door to door in other villages, performing these tricks for money. In other words, we found a village full of beggar Franz Hararys.

When we returned to the hotel, the ticket table was abandoned and Tripti was gone. I had no way to get in touch with her and cancel our illicit rendezvous.

So here I am, at 8:25 p.m. in Dhaka, sitting in my hotel room, waiting for Tripti to arrive, killing time by crapping out my intestines and researching acid attacks on Google. There are as many as 341 attacks in Bangladesh a year, most of which involve women. The weapon of choice is sulfuric acid, usually poured from a car battery into a cup and then thrown on the woman's face. The disfigurement that results is more hideous than anything I've seen in a horror film. And these women are the lucky ones. The unlucky ones are forced to drink the acid.

Of course, I could be horribly wrong about Ali Raj and his men. Perhaps they're actually on my side and protecting me from Tripti. Maybe they want to save me from a marriage trap she is laying.

Or maybe they're not actually cockblocking but hitting on me. According to one Website, five percent of Bangladesh's population is homosexual.

I wish she'd get here already. The Internet is a dangerous tool in the hands of a paranoid man with time to kill.

Five Google searches later, I hear footsteps in the hall. Getting closer. A knock. Why doesn't she just use the key I gave her?

I hear her voice. There's a man's voice, too. She's with someone. This is not a good sign.

"Be right there!"

I'm going to e-mail this to myself. Hopefully, someone will check my account and find it if anything happens to me. Maybe I should copy Bernard just in case.

Wish me luck. Or don't. I probably deserve whatever's coming to me.

... AND THEN ...

SO YOU MADE IT BACK IN ONE PIECE.
HEY BERNARD.
I GOT YOUR E-MAIL.
DO YOU THINK I'M CRAZY?
NO, I'M USED TO IT.
DID YOU FINISH YOUR SKETCHBOOK?
I'M FOUR PAGES AWAY.
SO, TELL ME WHAT HAPPENED...

"...WHO WAS SHE WITH THAT NIGHT?"

THAT'S CREEPY. SO WHAT DID YOU DO?
WE JUST MADE AWKWARD SMALL TALK. THEN HE MADE HER LEAVE.
AND THAT WAS IT?
NOT EXACTLY.
THIS IS FOR YOU.
HOTEL
Hi Neil how
angery. But i
T-shart and hard bealt
good girl this. but i like very
and i love you. i send your
love and kiss
Tripti
MAGIC
TOUR
WHAT'S THAT?
A NOTE FROM A WOMAN.
CAN I SEE?
SURE.

Hi Juan how you?
WHAT DOES *CHULA* MEAN?
Chula
IT MEANS *CUTIE*. IT WAS MY PET NAME FOR HER.
SO SHE WAS SLEEPING WITH ONE OF THE MAGICIANS ALSO?
THEY ACTUALLY DIDN'T SLEEP TOGETHER EITHER.
I WAS SO BUSY LOOKING FOR REAL MAGIC.
I NEVER STOPPED TO THINK THAT SHE WAS A TRICK TOO.
SOUNDS LIKE YOU MET A FELLOW ***PLAYER***.
YEAH, DON'T BE FOOLED BY THE SARI.
WANT TO COME IN?
I CAN'T. HAVE TO FIND A STORY TO COMPLETE MY SKETCHBOOK.

RULE 4

KNOW THE TERRAIN BEFORE TAKING THE JOURNEY

MAGGIE

Maggie climbed, dripping, out of the backyard swimming pool, perfumed in gardenia and chlorine. The water pooled in small bulbs on the ridges of bone in her neck, the shelves of young muscle in her abdomen, the disappearing baby fat of her thighs.

She strode toward me, as fast as happiness, and I led her upstairs, my steps heavy on the white plush carpet. I was envious of the way she existed so completely and freely in each moment, and fought to clear the maelstrom of anxieties that circled my mind like wolves hunting a deer.

I flipped her onto the bed and, as she hit the mattress, a giggle dislodged, filling my bare white room with the sound of female. She lay there and waited for what she knew would come next. If I could just press my body tightly enough against hers, thrust myself deeply enough into her, slow my heartbeat enough to match hers, then I, too, could feel young and free and happy.

I'm not sure what she wanted from me, a man twelve years older, out of shape, and consumed by worry over another deadline in an endless series of deadlines. Perhaps she wanted acceptance, unaware that the need for it is not only insatiable but the cause of most mistakes made in life. Perhaps she wanted maturity, unaware that it's just a cage adults make children race toward so that they may one day be as miserable as them. Or perhaps she was so carefree that she didn't want anything except to give.

LINDA

Linda wiped away a snail track of sweat running down her temple, biting her lower lip for my benefit. She straddled me cautiously, her legs and arms tense against the bed to prevent full surrender. Her body was long and agile, like a ballerina's but with a woman's hips, and thick brown hair flowed over her flat curves, hiding a nakedness that still felt dirty to her. Her lips were swollen with kisses, her cheeks flushed with the hours of passion it took to get her to this point. Every particle of air in the sparse bedroom—the one she'd grown up in, cleared of childish reminders of who she'd been—was filled with her energy, her intensity, her nervous excitement. This was it.

"Go slow," she said. "Be gentle," she said. "Only maybe for a second," she said. Everything a girl would say after making the decision to have sex for the first time, she said.

And then she hesitated, like an orange bobbing on the branch one last time before breaking off its stem. Over the years, she had imagined this act in so many variations of scenery and colors of emotion, denying suitor after suitor who wanted to take it from her because they were like bounty hunters who wanted to put an outlaw in jail not to serve justice, but so they could claim the reward. It had to be just right, so that ten or twenty or thirty years later, she could call to mind every sensation and smile with the conviction that she'd done the right thing.

A giggle—nervous, childlike, womanly, awkward—escaped from her lips as she lifted herself and turned around decisively, sitting astride my bony hips and facing my feet. She set her gaze on a rectangular mirror atop the flimsy pine dresser that had loyally kept her secrets through every age, stage, and metamorphosis. She watched closely as she twisted her torso a little to the left, so that it arced like a model's, then focused her gaze on her face, so she could see what it would look like in the moment of surrender that she so carefully controlled. This was not about me; it was about her. And, in a slow second, charged with nineteen years of being a daughter and a sister and a child, it was done.

ME

And now I sit with them, Maggie on the left in summer dress, Linda on the right in suede skirt, both holding my hand, both thinking I will take them home tonight.

Their grips mirror their beliefs: Maggie's hand lies softly over mine, without worry or urgency, because she knows there will be plenty of time for intimacy later. But she is wrong. She is unaware that two feet away, the hand of her younger sister squeezes mine tightly, possessively, in tacit conspiracy. In her innocence, Maggie has allowed her conniving sister to accompany her on this date. And so the plot in the theater seats is thicker than that on the screen. Two sisters torn apart by a worthless man. And just like Esau and Jacob, Aaron and Moses, Bart and Lisa, the younger must win. That is the way of things.

And I, who thought I was the great seducer, who boasted of sleeping with model sisters, who validated himself in their embrace like a vampire drinking youth, was nothing more than a doll in their playset.

"We connected right away on a very deep level," Linda had told me that first night in bed. "But then Maggie threw herself at you, so I was just like, whatever."

But perhaps we'd never connected until Maggie claimed me. Perhaps, like me, Linda envied Maggie's freedom and spontaneity, and wanted to take away something of her older sister's. Perhaps she'd decided, on a subconscious level, to lose her virginity with the worst of intentions. And then, with love in her heart, with a smile on her face, with innocence in her eyes, she could once more make her sister feel like the black sheep. Perhaps waiting so long to lose her virginity was never a moral choice for herself, but one intended to make her sister seem like a slut in comparison.

The weapon of the youngest is never physical strength but emotional cunning. And now I am complicit in this trap. I must play my role: Maggie has slept with twenty-six men; I am just a footnote in her sexual history. But I am Linda's entire sexual history and its caretaker. I must keep her memory of the moment preserved in a bell glass. If it shatters, and one shard punctures her heart, the damage will be permanent. She is too smart: She chose the right man, one cursed with a conscience, which dictates that I not ruin her—or any woman—for other men.

And so I have no choice. Someone is going to get hurt tonight, and better the happy slut than the melancholy prude.

Maggie will never forgive me for this, nor will she ever forgive Linda. As I lie in her younger sister's bed that night, Maggie consoles herself with an ex-boyfriend.

A month later, with love in her heart, a smile on her face, and innocence in her eyes, Linda tells me—the one-man army she has used to stage her coup—that Maggie has moved in with him. Three months later, he has gotten Maggie hooked on crystal meth. A year later, Maggie has broken up with him for abusing her. Two and a half years later, Maggie is no longer recognizable as the carefree youth who once climbed dripping out of my swimming pool. She has married him. And, like air bubbles trapped in cement, the decisions we make in a moment haunt us for the rest of our lives.

RULE 5
WHAT YOU PERCEIVE IS WHO YOU ARE

She said she would pick me up in an old car.

"You'll hear it before you see it." Apologetically.

It was the first time I'd fallen in love with a car.

It was from 1972 and looked worse for the wear. The surface was pocked with small dents, dings, and patches of primer; the bumpers were rusty and looked like they'd seen a lot of action in their day; and the leather interior was torn up from years of constant use and neglect.

But its body was beautiful. It was sinuous and curvy, without a single flat edge; its front tire wells arched smoothly above the surface on either side, sloping into a hood so long you couldn't see the end of it from the passenger seat. When it glided out of the Phoenix airport, people turned their heads. It stood out from the other cars. It was magnificent, proud, unafraid of its defects because it knew its body shape compensated.

"This was the last year they made Corvettes like this," she said. "After 1972, they switched to plastic bumpers."

Her name was Leslie. And, though I'd never met her before, I was going to sleep with her. It was prearranged. Justin, one of my students, had offered me his cousin as a birthday present. It was above and beyond the call of duty. Normally I wouldn't have taken him up on such a creepy proposal, but he promised me that she wasn't just a lay. She was an education.

"She's been studying Tantric sexuality half her life," he said. "And she's discovered a G-spot in the back of her throat."

"That's kind of interesting," I replied, meaning weird. "How does that work exactly? Am I supposed to stick my finger down her throat and massage it?"

"No, something else." He smiled. "She's like a deep throat expert. She can take it all the way in, and work her throat muscles to make you experience something you've never felt before. This is next-level shit."

I was interested, in the classic sense of the word.

A newspaper columnist named Fanny Fern coined the expression that the way to a man's heart is through his stomach, proving to the world just how little women know about men. We can always go out to eat. But if a woman wants to make an impression that we'll never forget, even when we're eighty and on our deathbed and thinking about the two moments that made life worth living, all she has to do is give us the most masterful blow job of our lives. If she even hints that she's great at it, we'll chase her all night. Then, if she actually delivers, she'll never have to worry about a phone call the next day.

It's funny how much time women spend trying to figure us out when we're so simple. I think what's complicated is accepting how simple we actually are.

As Justin pitched me on his cousin, I thought about all the people in my lifetime who had promised to get me laid and never delivered. I remembered Marilyn Manson's bodyguard telling me he had two girls in his hotel room giving a sex show, but because he was married and couldn't sleep with them, he'd send them to me. I lay expectant in my hotel bed for hours, fresh from the shower, trying to stay awake in case sleep turned my breath bad, waiting for the knock. But the knock never came.

Only I came. Alone. Again.

So before my next trip to Phoenix, just to be safe, I called a thin, buxom Iranian girl named Farah, with heavy-lidded, glittering brown eyes. I'd met her last time I was in Phoenix and she mentioned buying a book on Tantric sex. This way, I figured, the Tantra thing would happen one way or the other.

"Yeah, I'm living with my father for now in Sedona," Leslie gabbed as we drove to the James Hotel. "I stay with my sponsor sometimes in Scottsdale, but he's been an asshole lately."

I wanted to ask her what she meant by sponsor. Was he her mentor in a drug rehabilitation program? Her sugar daddy? A client of some sort?

But the question seemed inappropriate, as did all the others I wanted to ask. I wasn't sure yet if the sex thing was really on—if she had also been informed that she was going to deep throat me tonight—and didn't quite know how to confirm the appointment.

Leslie wasn't the type of girl I normally slept with, or even talked to. Experienced would be a polite way to describe her face, which was a weird shade of red—not from the sun, but from some style of makeup application I'd only seen used by bag ladies on public buses. She had teeny teeth pressed close together, which would have been cute if they weren't out of proportion to her broad face, sabotaging every smile.

Her body, however, was glorious. She was a big girl. Not fat, but solid. Mighty would be a better word. Her pink-powdered breasts heaved out of her dress, daring you not to look at them. Her thighs were thick and muscular, and looked like they could perform all sorts of functions on construction sites. And her posture screamed sexuality and multiple orgasms. You could tell by the way her back arched away from the seat and thrust the full force of her tremendous chest into the steering wheel.

This was all so exotic to me. Though I tell girls I weigh 140 pounds, I've actually never been able to get above 126, no matter how much I eat or work out. Until recently, I had only dated really small women with low self-esteem, because that was all I could handle. This girl was an Amazon, a really trashy one, possibly even a real-life whore. It doesn't get any worse than that. And worse is what I'm all about.

When we arrived at the hotel, she reached behind her seat, grabbed a small overnight bag, and brought it with her into the hotel. As soon as I saw this, I knew Justin had made good on his promise.

I just had one major concern left.

"So, what are you doing for work these days?" I casually asked during dinner.

"I used to be a dancer," she said, "but now I'm between jobs."

As we talked further, I tried to pull more details from her. The best I could gather was that she'd been a stripper for six years, made a few adult films, and now used certain former clients for shelter, gifts, and travel. I suppose that makes her a prostitute, just as much as it makes any woman who dates or marries for money one.

After dinner, we took the elevator to my room. There still hadn't been a word or gesture of intimacy between us. Even though she was doing this for blood and not for money, there was something unsettling about the whole arrangement. Some guys enjoy having sex as a transaction, rather than an act of passion. But I get my rocks off as much through connection and, on a shallower level, validation as through the friction of flesh. I need to know that the woman I'm with wants to be with me because she genuinely likes me as a person—whether it takes three minutes or three years for her to come to that decision—or else the mutual surrender so key to the transgressive pleasure of sex never happens.

I decided to take some time to connect with her before the deep throating commenced.

"If you had to choose one thing in the world that makes life worth living, what would it be?" I asked as we walked in the room.

"Hmm," she said, nodding her head and pulling off her dress. Still thinking, she unhooked her bra. Her breasts were gargantuan. I could have placed a dictionary between them and they'd hold it like bookends.

She knelt in front of me and began unbuckling my belt.

We can always connect afterward, I decided.

"Why don't you stand in front of the bed?" she suggested as I stepped out of my pants.

I complied, as if following a nurse's instructions for a physical. She climbed onto the bed, rolled over, and dropped her head backward over the edge of the bed. I realized that this must be her special trick.

I stood in front of her and approached her open mouth with my dick in the air. It felt like some sort of carnival game.

She brought her hands up, wrapped them around me, and nudged me into her. Then she began adjusting her head in small movements, guiding me into her throat like a maze, until her mouth was at my base.

Euphoria swept through my body. In that moment, I knew my answer to the question I'd asked when we walked in the room.

She began sliding me back and forth inside her, slowly at first, clamping her throat and lips around me every time she hit bottom. Glancing down, all I could see were her outstretched neck and chin and, for some reason, they reminded me of the belly of a penguin. It was solely due to this image that I was able to refrain from orgasm and proceed to intercourse.

"I want to bring a girl out with us tomorrow," Leslie said, greedily puffing on a cigarette afterward. "She's got a gorgeous body. I've been trying to get with her for years. Maybe you can help me out."

My uncle used to warn me, "When pigs become hogs, they get slaughtered." I was about to ignore his advice and try to arrange a foursome.

"That would be cool," I told Leslie. "I was actually thinking about bringing along this Iranian girl I know who wants to learn Tantric sex. I told her you were a guru, so maybe you can show her a few things after dinner."

"Or during dinner." She smiled, exposing her teeny teeth. I couldn't imagine a weirder partner in crime. I was actually starting to like her, which was a good thing, considering that I'd just slept with her.

The following evening, after Leslie and I finished another game of penguin, there was a light, rapid knocking on the door. I opened it to find a woman with long legs encased in tight jeans, a flat, exposed abdomen, and a half-shirt clinging to large natural breasts.

Her face, however, was etched in permanent frown lines, stamped with dark circles around the eyes, framed in an explosion of frantic black hair, and crowned by a halo of drama. This was Samantha.

The first words out of her mouth were: "I need to borrow your phone."

Leslie's friend, Leslie's problem.

She took Leslie's phone, shut herself in the bathroom, and yelled at someone's answering machine as the bellhop arrived with three black bags. Samantha was moving in.

I left the room for the temporary refuge of the lobby and called Farah to warn her that my friends were going to be a little unusual. When I returned, Leslie was wearing a leopard-print dress with a plunging neckline and Samantha had changed into an imitation fur vest with nothing underneath.

When we walked through the lobby, a skinny bald guy sandwiched between two curvy giants dressed like eighties streetwalkers, every head turned. For a moment, I thought this was all a practical joke Justin was playing on me, but he's too broke to hire girls. Just to be safe, on the cab ride to the restaurant, I checked Leslie's ID to make sure she shared Justin's last name. Fortunately, her credentials checked out.

"I lost my credit card," Samantha prattled. "Do you guys mind if I borrow money just for tonight?"

"You're on your own, kid," I told her. I wasn't going to let her put me in the daddy role. If she wanted respect, she'd have to earn it.

Farah was waiting for us at the restaurant in a black strapless evening dress. She far outclassed my company.

"This is Leslie, the Tantra teacher I told you about," I said.

Farah smiled and greeted her. Only a slight, involuntary furrow down the center of her forehead gave away her befuddlement as to how this pink-boobed leopard woman could possibly be a spiritual guru.

The maître d' led us to a table in the outdoor garden, where a movie was being projected onto the wall. Conveniently, the film was *Last Tango in Paris*.

To break the ice, I ordered a bottle of wine and performed a few illusions I'd recently learned, including one where I cause a ball of paper to rise off the table and float into the air.

"If he can send his energy to objects, imagine what he can do with parts of your body," Leslie told Farah. She was a great wingman.

"That stuff scares me," Samantha interjected. Every word out of her mouth was a plea for sympathy. "I need more wine. Can someone get the waiter over here? I think I'm getting a migraine."

The meal was interminable. No matter what subject we discussed, Samantha managed to bring it back to her neuroses. If we were talking about the movie on the wall, she complained that her cable was out and the repairman wouldn't come over. If we were discussing sex, she complained that the guy she was dating hadn't called her all week. If we were exchanging stories about nights out in London, she went on a tirade against her brother because he's a travel agent and never gets her deals.

My head ached just listening to her. "Do you see a pattern?" I finally snapped. "Your repairman won't come over, your boyfriend doesn't call you, and your brother doesn't help you out. Maybe the problem isn't everybody else; maybe it's you."

Her face scrunched, her eyes puffed, and she fell quiet for the remainder of the meal. I could tell that she was adding the comment to her archive of victim stories to tell for sympathy.

I'd just destroyed the night's foursome. And I was fine with that. It wasn't worth the headache. After dinner, I told Leslie and Samantha that I was going to a party with the Iranian princess. They seemed fine with that, and said they were going to a dance club.

However, between the magic tricks I'd performed, which led Farah to think I had actual shamanistic powers, and the company I kept, which led her to think I had a perverted sex life, she kept her guard up. When she dropped me off at the hotel after the party, we made out tepidly in the car. She seemed to be accepting my kisses, rather than returning them.

I walked to the elevator, dejected. My foursome had turned into just me, alone, again. My uncle was right. When pigs become hogs, they get slaughtered.

When I stepped off the elevator, I saw Leslie, Samantha, and a third girl I didn't recognize smoking in the hallway and waiting to get in the room. I'd assumed they'd be out partying all night.

Their friend introduced herself as Dee. She was petite, with a quiet confidence and braided hair extensions that ran halfway down her body. Her skin seemed Latin American, her facial features Native American, her backside African American.

Inside the room, Dee pulled a water bottle out of her purse, took a sip, and handed it to Leslie. Leslie took a small swig, then handed it to me.

"GHB," Samantha warned.

I passed it back to Leslie unsipped. I officially owed Samantha one.

Leslie fished into her overnight bag and produced a metallic green dress with an oval cutout running from just below the neck to the navel. "Hey, you have to try this on," she said to Samantha. I admired Leslie's talents as an instigator.

Samantha emerged from the bathroom moments later, looking like a Christmas tree with a misshapen star. "This one's perfect for you, Dee," Leslie said, pulling a white mesh minidress out of her bag.

Dee did not use the bathroom. She pulled off her jeans and tank top, revealing a body designed for the covers of muscle car magazines, and put on the dress.

"Mmm, you look good," Leslie purred. She walked up to Dee, laid a hand on the center of her chest, and began making out with her.

I was in the presence of a professional.

Within minutes, Dee was spread-eagled on the bed with her dress hiked up and Leslie's face between her legs. I sat next to them in my dinner clothes, not on GHB, thinking, This is cool.

When I joined them, via the nearest available breast, Leslie looked up at me, chin wet, and grinned from ear to ear. She reminded me of a coyote eating carrion.

"It's too hot in here," Samantha said suddenly. "I need some air."

By air, she meant attention. "Come join us," Leslie trilled, rising from the bed to bring Samantha into the mix.

"I want to clean the room a little first. You guys go on ahead. Don't mind me." The room wasn't even messy.

"Maybe I'll join you guys later," she added awkwardly, unconvincingly. "Looks like fun."

Leslie returned to the bed and pulled my clothes off. She and Dee both went down on me.

"Do you think there's an ironing board anywhere?" Samantha asked.

This was becoming even stranger than a foursome.

"You know what you can do?" I suggested, once again ignoring my uncle's advice. "Grab my camera off the table and take some photos."

Leslie and Dee didn't object; there probably wasn't much they'd object to. As the flashes went off, and the two of them earned their way into my shortlist of deathbed memories, I tried not to orgasm. A woman's sexual appetite, once unleashed, is much more voracious than a man's, and if I blew it now, I'd be stuck on the sidelines for the rest of the game.

"What button do you press to see the photos?"

I ignored her. This was my moment to shine.

"I'm bored," Samantha moaned. "I'm going to take a bath."

Leslie jumped up. "I'll help you."

Samantha was doing this on purpose.

Ten minutes later, Leslie returned from the bathroom, rebuked, and asked me to take a shot.

I grabbed a towel, wrapped it around myself, and sat on the edge of the bath.

Samantha was sitting naked in shallow water, her legs bowed out like a bratty child's.

"Everything okay?" I asked.

"Yeah, I'm okay. I like it here."

I decided to push my luck. It is my nature to push my luck. I am a hog.

I slipped off the towel and joined her in the bath. As we talked, I massaged her arms and legs. She didn't stop me.

I circled my fingers around her nipples until they hardened, then ran my tongue across them. She didn't stop me.

I moved my hand up her leg, until it reached the apex, and traced my finger slowly down her opening. She stopped me.

"No," she said, pushing my hand away. "Too much."

I'd been so worked up from the activity in the bedroom that I'd neglected to turn her on enough. And that was fine. Two birds in the bed, I decided, are better than one in the bathtub. I'd have to share that aphorism with my uncle next time I saw him.

When I returned, Dee was going down on Leslie. I joined her, and eased my finger up to her G-spot. This was more like it.

Leslie moaned and arched her back. She shuddered to orgasm, then begged us to keep going. Dee and I switched positions, and Leslie quaked again. She begged for more. For what seemed like forty-five minutes, she kept us down there, giving her orgasm after orgasm. My jaw ached, my wrist hurt, I began thinking about how good a Caesar salad with huge seasoned croutons would taste. Leslie kept arching her back, making us work harder and harder for each orgasm. But, as greedy as she was, I didn't stop. I wanted to show my appreciation for what she'd arranged tonight.

"Wow, that bath felt so nice." The fun-ruiner had returned. "Do you guys mind if I call room service? I'm hungry."

"No," I told her. The last thing we needed was room service busting in on the action.

"No, you don't mind or no, I shouldn't do it?"

"No, now would be a bad time."

Leslie, somehow, managed to have another orgasm during all this.

"I'm just going to make some tea."

I don't care.

I put on a condom, made sure it was unrolled to the very bottom, then entered Dee while she was going down on Leslie.

"Oh, here's the ironing board."

She must be on crystal meth.

"Do you mind if I iron your shirt?"

I may be all about worse, but this was becoming a nightmare. It was like having sex with my mother in the room.

Eventually, both Samantha and Dee were satisfied and they fell asleep. Not even a thank-you.

"You can go to bed now," I told Samantha. "You're safe."

"That's okay," she said, sitting in the desk chair. "I'm an insomniac."

Definitely meth.

With my mind and heart still racing from the night's adventure, I had trouble falling asleep. Samantha, conscious of this, began reciting her life story—her father shooting himself in front of the family at a dinner party; her mother leaving her at an aunt's house and never coming back; her first love beating her throughout the ten years they dated.

No wonder she was always begging for help and attention: Everyone she loved had left her or abused her. And, decades later, she was still searching for the safety she'd never felt as a child. Thanks to the needy way she went about it, however, she ended up replaying her childhood rejections with every new person she met instead.

I actually began to feel bad for her. Then I fell asleep.

In the morning, I woke to the sensation of Dee biting my neck. We were the only ones in the bed. It felt kind of empty.

"Where's everyone else?"

"They're in the bathroom," she whispered.

She reached around and stroked me. "Do you have another condom?" she asked.

I put one on. She rolled onto her side, with her back to me, and I entered her. When I began to moan, she whispered for me to be quiet, as if worried Leslie would hear us. I couldn't understand why this was an issue. Maybe she thought I was Leslie's man. Maybe we were breaking some unwritten law of the ménage à trois. Or maybe she'd just forgotten to bring her dildo that morning.

An hour later, we packed our bags, left the room, and took the walk of shame through the busy hotel lobby. Samantha offered to drive me to the airport and, as the four of us waited for her car at the valet stand, she grabbed my hand.

"Your skin is so soft," she said coquettishly. This was so out of her character that I didn't know how to respond.

Her car was not old and sleek like Leslie's. Just a beat-up white Malibu from the nineties. Its dented body, grinding brakes, neglected interior, and broken taillight conveyed nothing but hard living and bad luck.

After she pulled up to the terminal, Samantha applied lipstick, pulled an envelope out of her purse, and covered it with kisses. Then she handed it to me. I took a last look at the women in the car. I was actually going to miss them.

I guess I had connected with Leslie after all—and, as much as I was loath to admit it, with Samantha, as well.

As I flew back to the relative normalcy of home, I opened the envelope. Inside was a torn scrap of paper covered front and back with tiny scrawl:

> *Please call me next week or e-mail me. You turned me on very much, and I haven't felt what you had me feeling in a long time. It was a relaxing, sexual feeling. A turn-on that I never felt. I would have liked to experience being with you! I think you're a wonderful guy. I want to thank you for making me feel the way you did, and you didn't even know that you did. I sure wanted to suck your dick.*

The next day, I loaded the photos she had taken onto my computer. They were the most compromising images I'd ever been in: I could actually see Leslie's insides for several layers. It would be a disaster if they ever leaked on the Internet.

I opened a secure deletion program to wipe them off my computer forever. And then I sat there, listening to my hard drive grind out 0s and 1s, until the night never existed. They were from another world. And I had fit into that world a little too well.

RULE 6
EXPECT THE BEST, PREPARE FOR THE WORST

Dear Stacy,

You write the best e-mails. They are so thoughtful, warm, and tender. I wonder sometimes what it would be like to kiss you. I imagine that you would fully give yourself with a kiss, that it would be, like your e-mails, thoughtful and tender. I think of the warmth of your mouth, the joy of the first intimate touch, and how at first you might be a little nervous, but as you relax into the feeling, you would get lost in the moment, and our bodies, time, and the rest of the world would just melt away into that one single kiss.

Good night, Stacy. I hope all is well.

—Neil

P.S. I was pleased to hear that John and your sister are engaged. Please pass on my congratulations, and my gratitude to them for introducing us.

Dear Neil,

Your description of our kiss leaves me rather speechless. I can definitely feel the nervousness at first, but then the love pours in as we embrace. I don't want to sound corny, but that simply is how I envision our kiss: like the sun, love just warms everything about us.

I should warn you of something, though: I am a novice when it comes to kissing and sexuality in general.

Here's the short version of the story: for many years, I have battled anorexia nervosa, and because of my low weight over a period of time, my sexual experience remained at zero. Only recently have I begun to branch out and respond to sexual stimuli, which makes me a late bloomer at twenty-eight.

The next time I see you, I may be a bit heavier than I was in Chicago. I seem to have overcome the disease in the last few months. Well, not completely, but let's just say I've eaten a lot of chocolate chip cookies lately!

So, I do not mean to shock you, but that is my story. I am a very loving person, and I have so much love to give, but my knowledge of love-making is about minus ten. But wouldn't it be fun to learn, and start with the most beautiful kiss of the century?

When can we see each other and fulfill the wish? I can surely swing a visit to L.A., but only if you're willing to have me after all I've divulged in this message.

Keep enjoying yourself and write back soon.

Yours,
Stacy

Dear Stacy,

I'm writing this from Australia. I arrived safely yesterday, and wanted to thank you as soon as possible for sharing your story with me.

I don't want to make you wait, wondering what I'm thinking. So I will let you know now that I truly appreciate your candor and honesty. I would never think any differently about you as long as you are making progress. So you can put those worries to rest. I promise to be a patient teacher. If you're a really good girl, I'll even buy you some chocolate chip cookies.

I remain willing and eager to have you visit, and see all the places I've been telling you about. How does February 21 to 24 work for you?

E-mail me your address, and I'll send you a postcard and show you the beach on the Gold Coast where I surfed today. I miss you, too. Funny, huh, considering that we've only spent a total of ninety minutes together?

—Neil

Dear Neil,

I really have no special reason to write: just wanted to chitter-chat with you since I am so exceptionally fond of you (on some level, let's face it: I love you). Right now I am looking at icicles the size of lances hanging off the eaves of our roof, and I am thinking of you on the Gold Coast surrounded by gold. Gold: the alchemy that we create, you and I, together.

Send me messages—messages full of your joy and love and whatever you have to spare. If you need to vent, put it here. If you need to wax ebullient, put it here. If you need to say a cuss word, put it here. If you need anything, put it here. You are guaranteed a reception and a proper response. Just because I care so deeply for you.

In the meantime, just know that my crush keeps getting bigger every day. By the time I visit you on the 21st, I'll have pummeled you into the ground with my crushing affection. Hope you don't mind!

Love,
Stacy

Dear Stacy,

Apologies for the delay. Thank you again for another beautiful e-mail. I look forward to your visit, and want to assure you that I have no expectations of you or for anything to happen, just like I hope that you have no expectations of me. I must admit that I worry about your crush: I hope that I can live up to it. Looking forward to next week. Expect to see me waiting for you at the baggage claim. I'll be the one carrying the tray of chocolate chip cookies.

—Neil

Dear Neil,

Thank you for a lovely trip to Los Angeles. I had an unforgettable time exploring the Getty Museum with you, and it was a thrill learning to surf.

While I am disappointed that things didn't work out for us, I will savor forever the alchemy of our kisses and my first sexual explorations.

I am of course aware that gradually you distanced yourself from me, and I apologize for my lack of sexual experience and my crushing affection and everything else that probably scared you away. Because of my condition, I am not as comfortable with myself as I'd like to be.

I think you are a special person, and I will always have a space in my heart for you. Thank you again for showing me your world.

I am sad, but I will pray for you.

Love,
Stacy

Dear Stacy,

It was great to see you. And I feel the same way. You write the most beautiful e-mails I've ever received, and I will treasure them always.

I suppose an explanation is in order: I was so excited to see you at the airport, after all our e-mails, each one increasing in intensity. And, I must admit, at some point, I was a little scared, as well. When we went back to my house, I think reality set in. When I discovered that you still had your hymen, I realized you were no ordinary girl and this was no ordinary experience.

I didn't know if I could live up to your expectations, or ever reciprocate the immense reservoir of feeling you have for me. So I thought it would be better to back off and be friends, and let you have that other experience with the incredible person you're really supposed to be with. I can be a great lover, but I've always been a horrible love. I don't know if it's an emotional failing of mine, or if it's simply that our worlds are so different. You go to church every Sunday; I write books on Marilyn Manson.

You have so much love in your heart and goodness in your soul, and I'm glad that you were able to share just a little of it with me.

Are you familiar with Ryokan's poetry? The first part is by Ryokan and the second part is by Teishin. These are what I call good for the night poems.

Ryokan's letter:
Having met you thus
For the first time in my life,
I still cannot help
Thinking it but a sweet dream
Lasting yet in my dark heart.

Teishin's reply:
In the dreamy world,
Dreaming, we talk about dreams.
Thus we seldom know
Which is, and is not, dreaming.
Let us, then, dream as we must.

Good Night Stacy,
Neil

RULE 7
WHATEVER'S IN THE WAY *IS* THE WAY

"I was at a friend of mine's house and this storm came up out of nowhere, man, with big clouds that looked like snakes standing up," he was saying, his deep voice reverberating off the hotel room walls. "I had one of those little twelve-dollar cameras in the glove compartment of my truck and I just snapped pictures. When I got the photos back, there was an image of God with his beard blowing in the wind, standing up in the storm."

He was one of the most important musicians of the century. After weeks of work, I had finally persuaded him to sit down for a two-hour interview. And everything was going well—until the last ten minutes. That was when his granddaughter walked in the room. Suddenly, I found myself unable to focus on a word he said.

She had thick black hair, long muscular legs, a high forehead, and tremendous breasts lifted high in her sweater. Her silhouette was the kind people made stencils out of and stuck on the mud flaps of trucks. Judging by her proud posture and haughty air, she seemed well aware of the effect she had on men. But, worst of all, she seemed bored.

She lounged on the bed, picking feathers out of the pillowcase. In her mind, I was just another white guy pumping her grandfather for trivia from fifty years ago.

I had to do something to change that.

"In my belief, there's a supreme being who can show himself whenever he feels like it. But he comes angry at the way we live and treat one another. He didn't mean for us to fight like cats and dogs. He meant for us to get along and love one another until death takes us away," he concluded.

"Let me ask you a question, since you understand human nature so well," I began. I needed to pull his granddaughter into the conversation: "You can help out, too, if you want."

She glanced up indolently, mildly interested. "You know how they say women are more attracted to power and status than looks?" I continued, beginning an admittedly ridiculous opener I'd been testing lately to start conversations with women. "I was talking to a friend about it the other day and he asked a good question: 'Then why is it that most women would rather sleep with Tommy Lee than George Bush? Isn't George Bush one of the most powerful men in the world?' "

"Who's Tommy Lee?" he asked.

"He's the heavy metal drummer who did that sex tape with Pamela Anderson," his granddaughter explained.

"Well, that tells you something right there," he said. "It's because rock 'n' roll is soulful. You listen to it to get away from all that political bullshit."

"George Bush is ugly," the granddaughter opined, too beautiful to bother with the actual point of the question. "That's why no one wants to sleep with him."

Weak answers to a weak opener, but it had served its purpose: the focus of the conversation had now shifted to her.

"She wants to move here and model," he explained. "She's not like them toothpick girls. Skin and bones do not excite me. They need young girls with figures like Alicia's."

He wrestled his pocket for a mint, then shoved it in his mouth. "The burning went to the wrong place," he coughed.

This seemed to remind her that he was old and that time was short. She massaged his shoulders, waited for him to regain his composure, then made her agenda known: "Don't forget, you promised to take me shopping."

"This is her first time in New York," he went on, "but I reckon I'll be sorry I brought her."

These were all clues: model, shopping, new to city, Grandpa's reluctance to shop. Before I put these clues to use, there was one thing I still needed to know. "You're how old now and you've never been here?"

"Twenty-one," she replied.

The word granddaughter had worried me.

"She has to go to Century 21," I said, planting the seed to spend more time with her. "They sell every designer brand you can think of for practically nothing. She'll spend hours there."

After the interview, he decided to take a nap. I gallantly offered to take Alicia off his hands and escort her to Century 21.

She glided by my side through the streets, speaking rarely, smiling never. This was her first time in New York—dense with noise, drama, dirt, culture, chaos, life—and she was sleepwalking through it all. She seemed to exist in a glass box that separated her from the rest of the world. And I wanted, more than anything, to smash through it.

I once told the story of Sleeping Beauty to a young cousin of mine. "How can a prince fall in love with a girl who's sleeping?" she asked afterward.

"Good point," I replied. "She may be beautiful, but they haven't even spoken. What if she's a complete bitch?"

This is probably why relatives don't allow me around their kids.

At the time, I didn't have an answer for her. Now I did: He loves her simply because he has the power to wake her.

At Century 21, I tried to flirt with Alicia, choosing the ugliest outfits and insisting she try them on. But no matter what I did, I couldn't break through her reserve. She still saw me as an antique collector rummaging through the closet of her grandfather's mind.

She left the store two hours later with a purple satin dress, a lace skirt, and an extra-large men's polo shirt. The shirt, she said, was for her boyfriend.

This complication would have been much easier to take if the shirt had been a size that was easier to compete with. Like extra-small.

That night, I had plans to see a stylist I was sleeping with named Emily. I'd talked to her for a few minutes at a party once. Afterward, she found my e-mail address online, wrote to me, and suggested getting together for coffee.

"You're like heroin," she said when I arrived, late from shopping with Alicia. "All my friends say to stay away from you because I'm starting to fall in love with you."

When she pulled me into the bedroom and began undressing me, I imagined that her hands were Alicia's hands; I saw Alicia's mouth wrapped around me; I grabbed Alicia's thick black hair.

I had sex with Emily three times that night, and every time, I closed my eyes and imagined she was Alicia.

It was the most passionate sex Emily and I had ever had.

The following evening, after watching Alicia's grandfather perform, I went backstage to pay my respects and invite Alicia to a party at the Tribeca Grand Hotel that night. Slowly, languorously, as if she'd been asked to pass the sugar at the end of a long meal, she gave her consent: "Okay, pick me up at my hotel after I take Granddad back."

Because it was my last night in New York, and I didn't know whether or not Alicia would go out after the concert, I'd invited a date to the Tribeca Grand earlier that day. Her name was Roxanne. She was five foot two and one of the most sexual girls I knew.

An hour and a half after the show ended, Alicia emerged from her hotel, wearing the tight purple dress she'd bought. The cabdriver, the students across the street, some guy riding past on a bicycle all did a double take.

"I had to talk to my boyfriend," she said, apologizing for her tardiness. "We haven't spoken in like a week. He's so boring."

Sleeping Beauty was mine again to wake. Suddenly, extra-large meant nothing to me.

Roxanne was waiting for us in the lobby of the Tribeca Grand, wearing a spaghetti-strap top that exposed her little-doll back. She hugged me tightly, peering up through heavy black mascara. There was something mischievous in her eyes, her smile, her carriage that communicated she was willing to try anything anytime.

I had met Roxanne at a concert last time I was in New York. She worked part-time as a model for illustrators and had appeared on everything from biscuit tins to sex-position guides. Her boyfriend played drums in the small local band we were watching. And she invited me to the afterparty at the singer's apartment.

Roxanne, her boyfriend, and I spent most of the party lying on the host's bed, while he sat in a chair nearby. As Roxanne and I talked, her boyfriend rose to his feet, walked into the front room, and dragged a very drunk blonde onto the bed with us. Within seconds, he was making out with her. Two minutes later, he had her naked.

Roxanne didn't seem to mind, chiefly because she was too busy flirting with me: unnecessary touching, unsubtle innuendoes, unmistakable body language. Hesitantly, I took the bait. I looked over her shoulder as we kissed to see if her boyfriend minded. He was already fingering the drunk girl.

This is typically a sign of an open relationship.

I began making out with Roxanne more intensely. She grabbed me through my corduroys as her boyfriend began fucking the drunk girl. Some sort of jewelry glinted off his dick, rattling with each thrust. It was at this point that the singer left his own room.

As we fooled around, Roxanne kept glancing over at her boyfriend. She seemed upset, not necessarily because he was having sex with someone else, but because he was being inconsiderate of her while he did it.

She pulled down my pants and gave me an aggressive blow job. Then she grabbed a condom from her purse, slammed herself on top of me, and tried to outfuck her boyfriend. She ground herself vigorously against me, stuck a finger in her ass, and moaned loud enough to wake the whole building. This seemed to be how they fought.

It wasn't a good experience, but nobody ever said all experiences had to be good. Sometimes they're just experiences.

They broke up a few months later and, now that Roxanne was single, I was looking forward to sleeping with her under normal circumstances if things didn't work out with Alicia. Every single man needs a sexually adventurous woman he can count on to distract him from the fact that he is unloved.

"I brought some Ecstasy," Roxanne said after buying the first round of drinks at the Tribeca Grand. She pulled an orange pill bottle out of her purse and dumped a white tablet into her hand.

I'm not a fan of psychedelic drugs, mainly because they last too long. The word trip is appropriate: Like an airplane ride, there is no way to get off until you land. More important, I didn't think hugging a speaker for six hours would improve my chances with Alicia.

Pinching her teeny fingers together, Roxanne cracked the pill in two. One half instantly crumbled to pieces in her hand. Without even asking if I wanted it, she lifted the hand full of Ecstasy dust, clamped it over my mouth, and dumped the contents inside.

I tried to keep my cool, but my eyes widened in horror, as if they'd just seen the devil. I needed to find a way to keep from tripping. I couldn't just start spitting all over the club. So for the next five minutes, I kept bringing my glass of Jack and Coke to my lips and, instead of taking a sip, casually drooled the contents of my mouth into it. Then I went to the bathroom and poured the drink into the toilet. For the next hour, I was on edge, paranoid that the pill had absorbed into my bloodstream anyway.

Then I noticed Roxanne giving Alicia a massage on a couch upstairs. She'd already gotten further than I had with Sleeping Beauty. And that was fine with me, because it meant two things: The first was that I had succeeded in expelling the Ecstasy, because she was clearly in a drug-induced, tactile state and I still felt normal. The second was that a change of plans was in order. I might not have to choose between Roxanne and Alicia after all.

"My friend Steven has a great loft where I'm staying," I told them when their rubdown ended. "He and his roommates usually have parties every night, so we should see what's going on."

Roxanne, Alicia, and I took a cab to Steven's house, detouring at a corner deli to buy supplies: a bottle of Cabernet, Sun Chips, and turkey sandwiches on stale bread.

Inside the loft, the party had long since ended. Not only were Steven and his roommates sleeping, but two other guys were crashed out on couches in the living room. Unfortunately, I didn't have my own room. I had been sleeping on a futon on the floor across from the couches.

Roxanne and I sat on the guest futon. Alicia took a seat at a breakfast table a few feet away, unwrapped a turkey sandwich, and casually began eating it. I admired her ability to remain unaffected no matter where she went and what she saw. However, I was running out of time. There had to be some way to break the glass box in case of emergency.

"Hey," I whispered to Alicia, trying not to wake the two guys sleeping on the couch. "I have to show you the coolest video before you go."

My best wingman is my laptop.

She walked to the futon and perched on the edge with her arms wrapped around her knees. I showed her a clip of a species of bird that actually moonwalks across tree branches. I probably oversold the video, but it served its purpose, getting her on the futon.

It was now time to kiss Sleeping Beauty. Otherwise, she would return to the hotel and actually go to sleep.

I told Alicia and Roxanne that I'd recently had an amazing experience where two masseuses worked on me at the same time, in perfect synchronization. This procedure was known as the dual-induction massage, and I'd used it many times to segue into a threesome.

First, Alicia and I gave Roxanne a massage. Then I took off my shirt and they massaged me. Finally, I told Alicia to lower the top of her dress and lie on her stomach.

Typically, during the dual-induction massage, the energy in the room begins to shift and the inevitability of a safe, fulfilling, three-way sexual experience begins to dawn on everyone.

But this time, there was no shift in energy. Rather than relaxing into the touch and the sexual possibilities, Alicia lay there and quietly accepted the massage. Running my hands down the smooth, broad expanse of her back was as satisfying as it was frustrating, like smelling fresh bread in a locked bakery. I began to worry that she was politely waiting for her opportunity to leave, thinking we were some kind of creepy swinger couple who did this all the time.

Afterward, Alicia rose off the futon, pulled her dress up, and went to the bathroom. She didn't seem happy. She didn't seem upset. She didn't seem much of anything.

At least I'd tried. I was fooling myself by thinking Roxanne and I were Prince and Princess Charming anyway; we were more like the villains she needed to be rescued from.

"What do you suppose Alicia's thinking right now?" I asked Roxanne.

"I have no idea."

"Let's just check out her vibe when she comes back from the bathroom. And if she's not down, we'll put her in a cab."

Alicia returned from the bathroom to her perch on the edge of the futon, as if waiting to be dismissed. I'd definitely pushed her too far.

"Well, you should get some sleep before your trip tomorrow, so let's find you a cab."

She laid down next to me, hugged me good-bye, and said, "Thanks."

In the moment she hugged me, I sensed it was on. The energy shift I'd been waiting for had occurred.

I raced toward her lips, worried that if I hesitated for even a second, she'd be out the door. She melted into me. I could feel the glass box heating and cracking beneath my touch, falling off her skin in large panes. Faint murmurs of pleasure bubbled up through her lips.

Roxanne lay on the bed behind me. I turned around, pulled her close, and made out with her. Then we began massaging and licking Alicia's breasts through her dress. Alicia lazily raised her arms, signaling that she was ready for it to be taken off.

Alicia was not a giver, but she was a great receiver. Her back arched and her hips flexed, showing off a body so perfect that all the owner had to do was possess it to be a good lover.

When I removed Alicia's panties, she was drenched. I ran to my suitcase, dug for a condom, and returned to the bed. I positioned both girls on their backs and entered Alicia as I made out with Roxanne. Then I entered Roxanne and made out with Alicia.

To my surprise, the girls didn't hesitate once, even though there were two guys sleeping—or pretending to sleep—on couches in full view of the action. One of my friends, when he's having sex with a beautiful woman, thinks, I deserve this. I kept thinking, I can't believe this is happening to me. Are they blind?

A swinger couple I know used to tell me about their threesomes and, with delight and wonder in his eyes, the man would talk about his favorite position: the triangle.

The time had come to experience the legendary triangle. I lay on my back, and told Alicia to ride me. Then I had Roxanne sit on my face, opposite Alicia, so the two of them could make out.

However, I never felt the cosmic sexual flow my friend used to talk about. Instead, I felt blind and smothered. Roxanne was sitting on my eyes.

Not that I'm complaining.

Afterward, Alicia spoke first. "That's the first time I ever did anything like that," she said quietly.

"You mean a threesome, or being with a girl?" I assumed she wasn't talking about the triangle.

"Both," she said.

"How do you feel?"

"It was . . ." She paused. ". . . good."

She was never much for words.

• • •

Alicia and I stayed in touch after that. We had long phone conversations, during which her glass walls continued to fall away, exposing a goofy personality and wry sense of humor.

"Grandad likes you," she said one night. "He wants you to come visit us at home."

A week later, I flew in to spend the weekend and continue the interview in a setting few journalists ever got to see. Alicia picked me up at the airport and we drove to his home.

"I don't do this for just anyone," he said in his barreling voice when I arrived.

During the day, I watched him work in the studio. That night, Alicia snuck into my bed.

The next morning, at 6, her grandfather burst in the room. He took a look at us cringing under the sheets, then said to her, "I knew you were black-topping Neil."

He let out a loud, playful laugh, then turned to me. "Come outside, I want to show you something."

I followed him through the house and out the door. We stood in the grass and he pointed to the dawn sky. "Right there," he said. "What do you see?"

"Clouds."

"Look closer, man. What do you see in the clouds?"

They looked like smoke puffs, but he seemed so excited I didn't want to let him down. "God?" I asked.

"Yeah, God," he said, pointing at a thick wisp of cloud extending high into the sky. "You can never tell what He has in store for you. He moves in mysterious ways."

"Yes," I told him. "He definitely does."

RULE 8
EMOTIONS ARE REASON ENOUGH

I've made a horrible mistake.

I got drunk and may have married someone the other night.

And now I'm worried I'll never see her again. Or maybe I'm worried that I will see her. I'm not sure which would be worse.

I don't know her age, where she lives, or her last name.

Well, I suppose I know her last name now.

I'm not the type to blame other people for my mistakes, but if I had to point a finger, it would be at Ragnar Kjartansson. All you need to know about him are two things: One, he's the singer in Iceland's only country band. Two, he's the first male ever to graduate from Husmadraskolinn, a school for housewives.

He is my tour guide here in Reykjavík, the capital of Iceland, and I don't mind saying that he's not a very good one.

The night in question began at Tveir Fiskar, which either means Two Fish or Three Raincoats, depending on what time of day you ask Ragnar. It's one of the only places where they serve whale steak and whale sushi in Iceland. They also serve rancid shark, which is best eaten in bite-size pieces and washed down by a shot of Black Death. The former tastes like belly-button lint, the latter like Windex.

"We must drink," Ragnar slurred, handing me my third shot of Black Death, "to being pathetic."

He had been on a bender for months, ever since his girlfriend, Disa, left him and took the TV. Without the TV to distract him, he explained, all he did was think about her.

"I should have married her," he went on, bobbing his head into mine. "You only get one chance at perfect love."

After dinner, as Ragnar struggled to pull a red wool sweater over his head, he suggested, "Let's go drinking."

"Isn't that what we've been doing all night?"

"That wasn't drinking. I'll show you drinking, the Iceland way."

Evidently, drinking the Iceland way meant vomiting under a table, urinating on a bus, getting in a fight with a teenager, and passing out in a crosswalk. Because that's exactly what Ragnar did over the course of the next three hours.

"Get up." I nudged him. It was October in the frozen north and he was wearing just a sweater. "You're going to die out here."

"Go on without me," he mumbled. "The bars of Reykjavík need you."

Even in his drunken stupor, he was trying to make me laugh. I hoisted him to his feet and brought him to the safety of the sidewalk. And that's when I saw the girl I would marry that night.

She was accompanied by some twenty tourists, all of whom were attending Iceland Airwaves, a music festival I was in town to write about. I recognized a photographer in the group and stopped to talk.

He introduced me to his friends. The only word I remembered was "Veronika."

She reminded me of the new wave singers I used to fantasize about in the eighties. She was petite, with spiky black hair, heavy blue eye shadow, laughing eyes, and full lips parted slightly to expose a perfect row of white. As soon as I saw her, I was smitten.

"Is he going to be okay?" she asked, gesturing to Ragnar.

"Yes, he's heartbroken."

"I wish my heartbreaks were like that."

"Yeah, he does look pretty happy for a guy who's lost his perfect love."

"I've never had perfect love," she said. "I wouldn't even know how to recognize it."

"You don't have to recognize it. You just know."

One of the things I've learned from traveling with rock bands—besides how to play FIFA World Cup soccer on a moving bus, survive without showering for seven days, and sleep inches away from five people who also haven't showered for seven days—is that groups move at the speed of their slowest member.

And, considering that most of Veronika's friends were drunk, they weren't going anywhere soon. So I suggested slipping away, finding something interesting to do, then rejoining them in a little while.

"What about Loverboy?" she asked, gesturing to Ragnar.

"He can be our third wheel. Every date needs one."

She looked at her friends, then smiled her consent. We backed away wordlessly, with Ragnar wobbling behind us.

"*It's hard to be loved*," he began singing. "*Baby, I'm unappreciative.*"

"No wonder she broke up with him." Veronika laughed. I liked her. In order to be alone with her, however, I'd have to dismiss my hapless tour guide. I knew he'd understand—or, more likely, forget. So I flagged a taxi and stuffed him inside.

As I closed the door, he grabbed the bottom of my jacket. "Don't say no to love," he slurred. "Or you will be pathetic like me."

"I feel bad for him," Veronika said as he sped away.

"Don't feel bad for him. Being pathetic is an art form to him. He comes from a very accomplished family, so he distinguishes himself by being hopeless at everything: the worst drunk, the worst country singer, the worst boyfriend, the worse housewife."

"I suppose there's a sort of dignity there," she said.

Downtown Reykjavík on a weekend night is a combat zone, with bottles smashing against walls, cars careening onto sidewalks, and hordes of drunk teenagers zigzagging the streets. There's no malevolence in the air, like after a rugby game in England, just an absence of control.

Veronika and I found refuge in a small line outside the door of an after-hours club. She was from the Czech Republic and had been living in New York City for the last year. That was all I managed to learn before a guy with an unbuttoned overcoat, spiky brown hair, and a smooth face ruddy from the cold staggered in line behind us. He had a backpack slung on one shoulder and a big alcoherent smile on his face.

"Okay, okay," he blurted, barreling into our conversation. "From where do you reside?"

"The States," I replied curtly.

"It is beautiful for spacious skies," he said earnestly, as if he had just spoken magic words that would win him the approval of any American. "And may I ask as to whether you are male friend and female friend?"

"We actually just got engaged tonight," I said, hoping that would extinguish any hope he had of hitting on Veronika.

"That is blessed news." He smiled sloppily. Most people in Reykjavík were nearly fluent in conversational English, but he spoke as if he'd learned the language from technical manuals, greeting cards, and parliamentary papers. "For what measure of time do you date?"

"Seven years," Veronika told him, playing along. "Can you believe it took him this long to step up? He's scared of commitment." Definitely a keeper.

"That's because she's always nagging me about the trash and the cigar smoking and my checkered past."

"I can help," the guy said. "I can help. My surname is Thor. And I will marry you in holy wedlock."

"That would be great," I told him. It seemed like the perfect opportunity to make a connection with Veronika.

"Okay, okay, I need ring for ceremony," Thor said. He swung his backpack under his shoulder and began digging through it. "You are sure?"

"It's my dream come true," Veronika said, sighing.

"Okay," Thor prattled on. "This will be okay." He scooped a bottle of vodka out of his backpack, unscrewed the cap, and worked furiously to remove the metal ring around the neck. It snapped apart.

"Wait, wait." Undeterred, he produced a cell phone from the bag and slid off a metal loop that appeared to be an empty key ring.

He seemed so intent, so determined, so excited. We enjoyed watching the show. It was as if he'd been sent by a higher power to keep us entertained and prevent the awkwardness that usually occurs when two people who like each other hang out for the first time.

He said something in Icelandic to two guys in line behind him and they moved into position on either side of him. Then he cleared his throat and began:

"Dearly beloved, we gather today under God and witnesses to join pleasing couple in bonds of holy matrimony, okay, okay. Pleasing couple, I forecast your happiness for infinity. Your love is like sun shining in morning. It makes light of world."

At first, I thought he was simply playing the clown to amuse us. But as he went on, he seemed to be struggling, with all the soberness and poetry he could muster, to make the moment meaningful.

After five more minutes of grandiloquent speech, he furtively pressed the key ring into my hands, then addressed me: "Do you take this woman to be your wife in holy wedlock? Do you guarantee to love, honor, and protect her until death parts you apart? Do you guarantee to love her and only her in wellness and in health, okay, okay?"

"Okay."

"Do you take this man to be your husband in marriage? Do you guarantee to do all the things I just speeched to him, okay, okay?"

"Okay."

"I now pronounce you man and wife," he intoned loudly. "You may kiss on the bride."

As Veronika and I made out, I welled up with gratitude to Thor, who was already busy pulling something else out of his backpack.

"I insist on pleasure of gifting you with first wedding gift, okay, okay," he said. He then handed us each a small crescent of chocolate wrapped in blue-and-silver foil and made another rambling, romantic speech full of okays.

We thanked him for the passion he had put into the ceremony. And he beamed, proud of himself, then reached again into his backpack and pulled out a pen and a notepad.

"Please give to me your mail address, okay, okay," he said.

We both complied, figuring that he wanted pen pals.

"Make sure you spell full names with correctness."

He folded the piece of paper and put it in his pocket, then nodded happily and announced: "I will send certificate of marriage in mail, okay, okay."

I blanched for a moment, then realized he probably just meant a greeting card. He'd definitely gotten carried away with the whole charade. "What do you mean?" I asked, just to make sure.

"I am priest, of course," he said, as if it had been obvious the whole time. "I have certification with church. It is okay. We accept all religions."

Veronika and I both looked at each other, the same thought running through our minds: What have we just done?

Yet, oddly, neither of us told him not to prepare the certificates. He was so proud of himself, like a child who's taken his first shit on a grown-up toilet, that we didn't want to disappoint him. If he really was a priest, which he kept insisting, then it was too late anyway.

Once inside the club, we bought our priest a beer in exchange for his services, then snuck away to make out in the upstairs lounge. It was the most romantic first date of my life—and hopefully not the last first date.

There was little point in hanging out at the club, since we had no interest in talking to anyone else, so we left to find more adventure.

When we turned the corner, we saw Veronika's friends still standing on the sidewalk, exactly where we'd left them. We talked to them for a few minutes, but the conversation was awkward. They'd been standing there, doing nothing, while we'd been through so much. Our lives had, quite possibly, completely changed. So, once more, we slipped away.

She placed her hand softly in mine and we walked to the Hotel Borg like a couple on honeymoon. Upstairs, we collapsed onto the bed. It seemed obvious where this all was leading.

So obvious that, for the first time all night, Veronika began to get nervous.

"I've had the best time," she said between kisses.

My heart raced. I felt the same way. She continued: "This night is just too perfect. It can't be real."

We kissed again. Then: "I have to go."

And then: "This is too much."

Finally: "I knew you were going to try to do this."

It was clear what was going on. The specter of sex had cast gender roles on us. I was a man, moving toward pleasure, and she was a woman, moving away from pain. The same fear men have of approaching women, most women have of going past the point of sexual no return with men.

And this is not just because of the biological repercussions—pregnancy, labor, childbirth, nursing—but because most women have at some point been hurt by a man. So, before they risk giving themselves over to powerful emotions they have little control over, they want to make sure they're with someone who is being honest with them, respects them, and can reciprocate what they have to give—whether for a night or a lifetime. What many women secretly want is to throw themselves into the fire when they feel love without getting burned, scarred, or hurt. However, until scientists invent an emotional condom, it is typically the role of the man to reassure her before, during, and after that she's making the right choice. Not with logic, but with feeling.

"Before you leave," I told Veronika, "I'd like to tell you a story."

The story is not my own. It is about a man and a woman who randomly pass each other on the street one day. Both immediately get the intuition that the other is the one-hundred-percent perfect person for them. And, through some miracle, they work up the courage to speak to each other.

They walk and talk for hours, and get along perfectly. But, gradually, a sliver of doubt creeps into their hearts. It seems too good to be true. So, to make sure they're really supposed to be together, they decide to part without exchanging contact information and let fate decide. If they run into each other again, then they will truly know that they are each other's one-hundred-percent perfect love and will marry on the spot.

A day passes, a week passes, a month passes, years pass—and they don't see each other. Eventually, they each date other people, who are not their true love. Many years later, they finally pass on the street again, but too much time has gone by and they don't recognize each other.

"You see," I told Veronika afterward, "the lovers were lucky that fate allowed them to find each other once. When they doubted their feelings, it was like tearing up a winning lottery ticket and waiting for another one just to make sure they were really meant to win."

Afterward, there was silence. The metaphor had sunk in. We spent the night together talking about nothing but enjoying every word, fooling around but not having actual sex. Now I was not only indebted to Thor for the marriage, I was indebted to the Japanese writer Haruki Murakami for the honeymoon.

In the morning, as I lay in a state of semiconsciousness, Veronika kissed me good-bye. Reykjavík is a small city and we were both attending the same concerts, so we promised to find each other the next night. I spent the afternoon daydreaming about her and about our unexpected connection.

That night, we went to Gaukar a Stong, one of Iceland's oldest pubs. As seemed to happen every night here, the strong alcohol, the hallucinatory music, the clear air, and the winsome populace seized hold of me, and I gave myself over to the adventure the city had in store for me.

It began as I was ordering another Egil beer. A woman's voice to my right asked, "Are you American?"

I turned around to see a lightly freckled girl with short platinum hair dressed in combat boots, torn stockings, and a black sweatshirt emblazoned with a silver lightning bolt.

The conversation quickly turned to stories of sexual adventures, and she began talking about an orgy she had recently experienced. It soon became clear that the intent of the story was not just to share but to arouse.

It worked.

As we made out at the bar, a woman tapped her on the shoulder. I pulled back to see Veronika standing there.

"I'm leaving the club now," she told the girl coldly. "You coming with?"

"Yes," the girl said, grabbing her purse off the counter. Then, to me: "My friend's usually not this rude. Sorry. Nice meeting you."

It all happened so fast and unexpectedly that I didn't have time to explain myself to Veronika. I had no idea she'd been in the bar the whole time, just as she had no idea I was there—until she saw me making out with her friend. I suppose there was nothing I could say to her anyway, other than she was right when she said that meeting me was too good to be true. I'd already hurt her.

And now I'm sitting on the flight from Reykjavík to Los Angeles, replaying every moment in my head. I have no idea how to find her—or if I'm actually married to her. All I have to remember her by is the blue-and-silver foil chocolate in the pocket of my jacket.

Days pass, weeks pass, months pass, and I never hear from her again. Yet I can't get her out of my mind. My allegory has backfired on me and I've somehow convinced myself that we're the living embodiment of the Haruki Murakami story.

I try to find her on MySpace, but there are too many Veronikas without profile pictures in New York. I track down the photographer who introduced us, but he doesn't know how to get in touch with her. And the promised marriage certificates never arrive, which is actually more a relief than a disappointment.

I keep the chocolate on my desk as a reminder of my guilt, of my susceptibility to my lower impulses, of the fact that it was I and not she who so recklessly tore up the lottery ticket we'd been given.

Then, one night a year later, on a trip to New York, I see her—my one-hundred-percent perfect girl. She is at Barramundi on the Lower East Side, sitting at a table and drinking with friends.

The words "It's my wife" burst out of my mouth. The conversation at the table stops and everyone wheels around to face me.

"Hubby," she shouts, a wide smile breaking over her face.

I join them, and the hours pass. Eventually, it's just the two of us again.

I've dated many girls since meeting her. And she tells me she's in a serious relationship. Yet we still get along perfectly.

"I'm sorry," I finally say, "about, you know, making out with your friend. That was really stupid of me. I've regretted it every day since."

"You're just a man." She sighs.

"Does that mean my behavior is excusable because of my gender, or you're disappointed because I acted like a typical guy?"

"I guess both." I watch her lips sip her cranberry and vodka. "I should tell you that I had a boyfriend when we met."

"Is that the person you're seeing now?"

"Yes. But it's not perfect love."

"Then why do you stay with him?"

"I guess—" she pauses, reflects, decides "—because it's convenient love."

An hour later, we find ourselves at the apartment where I'm crashing. I show her the dead pet goldfish my host, Jen, keeps wrapped in Saran Wrap in her freezer, and then, tired and tipsy, we fall asleep on the sofa bed.

In the morning, we have sex for the first time. It is perfect. We fall back asleep afterward in each other's arms.

When I wake up, she is gone. I search the living room, kitchen, and bathroom for a note. There is none. Once again, I have no way to reach her. And I have a feeling that's the way she wants it.

The problem with one-hundred-percent perfect love is that sometimes it's inconvenient.

Back in Los Angeles a month later, I give in to temptation. I've been working all night and there's nothing to eat in the house. I peel the blue-and-silver foil off the wedding present Thor gave us. Small discolored flakes of chocolate drop to the ground. The candy has turned brittle from age, lost its shape, and faded from brown to inedible gray. There is no point in keeping it anymore. It will only attract bugs.

RULE 9
LOVE IS A WAVE, TRUST IS THE WATER

"I'm throwing up."

"Did you eat anything shady last night?" I ask her.

"No, I had what you did. How do you feel?"

"Fine, I guess."

"So."

This is where it begins to dawn on me that this is not a call for coddling. It is every unmarried man's nightmare—and many a married man's nightmare.

"Do you think you have food poisoning?" I ask. It's hard to just come out with the words. Their impact is too much to take.

"I don't know."

"Would you like me to get you some Emetrol?" I'm fishing now.

"Could you? Thanks." Pause. Wait for it. "And could you get a pregnancy test, too?"

When you know a slap across the face is coming, it actually hurts more.

I hang up the phone, brush my teeth, splash water on my face (an ex-girlfriend convinced me one morning that it's bad for the skin to use soap twice a day), and grab the car keys.

It is the worst trip a man has to make.

At the drugstore, I pick up crackers, ginger ale, and Emetrol antinausea medicine. Then I study the shelf of pregnancy tests. The E.P.T. Pregnancy Test seems the simplest: Pee on the white rod, then wait to see whether it displays a minus sign (indicating freedom) or a plus sign (indicating indentured servitude). I choose the kit with two test sticks. I may need a second opinion.

At the register, it is all too obvious what my errand is. This is far more embarrassing than buying condoms, though I imagine there are more humiliating things to buy. Like Preparation H. Or Valtrex. Or Vaseline and a plastic billy club.

They've probably seen it all.

I rush to Kathy's house. She answers the door wearing just a green T-shirt, her small face blanched, her blonde hair uncombed, her slender body beaded with perspiration. She looks great. No joke.

I unpack the groceries. The first thing she goes for is the ginger ale.

I carefully watch the pregnancy test to see if she's ready, but she just brings it into the bathroom with the medicine. Probably wants to wait. Too much to handle right now.

She doesn't mention it. Neither do I. She's already told me many times that she could never get an abortion. So there's no point in talking about it. Either we're screwed or we're not.

As she wanders around the house cleaning, I wonder how we're supposed to administer the test. The best thing would probably be to go into the bathroom together, as a unit. I'll stand by her side, politely averting my head while she pees on the stick. Then we'll lay it on the countertop and wait. We can run through what-if scenarios together then.

I suppose I could marry her. When we first started dating, I thought she was the one. People say you just know, and for the first time I did: I remember making out with her on the couch on our second date and thinking, I love this girl, and knowing I'd have to wait at least a month before I could actually tell her. I remember watching her sleep, and realizing that I would always love her, no matter how old and wrinkly she gets.

But lately she's been jealous. She doesn't like it when I talk to other women at parties, even though I make it plain to them she's my girlfriend. She doesn't like it when I answer my cell phone when I'm with her, even if it's the middle of a weekday, we've been together seventy-two hours straight, and it's a work call. And when we're lying together and she's looking into my eyes and, for a second, I remember that I have to take my clothes out of the dryer, there's hell to pay for thinking of anything that's not her. I can't live for the rest of my life with the thought police.

This test better be negative.

She shuffles to the TV and puts in a DVD of *Sex and the City*, season three. She's seen every episode at least a dozen times. Refers to them often.

She always tells me that she will love me forever, but how can love exist without trust?

The anxiety affects my bladder like beer and I head to the bathroom. While washing my hands afterward, I notice the pregnancy stick lying on the countertop. She's got it just sitting there, ready to go. That's kind of sweet.

I pick it up and examine it. I've never actually studied one before. There's a little minus sign in the indicator window.

First thought: She's not pregnant. What a relief.

Second thought: She took the test without me?

I walk out of the bathroom to find her lying on the floor in front of the TV where I left her. She's watching the episode where Charlotte and Trey decide to take time apart.

"Why didn't you tell me it was negative?"

She looks up at me and shrugs, "I didn't want to bother you."

Then she turns back to the TV. I know how the episode ends. I know how all of them end. They'll break up. Then they'll get back together again. Then they'll break up again. Some things just aren't meant to be.

RULE 10

THE COMFORT ZONE IS ENEMY TERRITORY

THE FIRST DAY

"Your balls are going to be in your throat and you'll be screaming in pain," she says.

"No," I tell her. "I can do it."

"Sure you don't want to wait a few more days?"

"I'll be fine. Now take off your pants."

Gina steps out of her pants and I lay her down on the couch. I want to make sure she's as close to orgasm as possible to make this easy on myself.

"No tricks, now," I warn as I enter her. "If I say stop, you have to stop."

It's different this way. I feel a sense of clarity I've never had during sex. My mind is alert and in the moment, instead of elaborately recording imagery to its fantasy database. I am detached from the friction and frisson, and as our grinding intensifies, my body begins to lighten and then dissolve.

She comes in slow, deep waves. Immediately afterward, she flails from side to side, as if the physical sensation is too much to take and she needs to crawl out of her skin until it subsides.

"I want to go surfing." These are the first words she says when she comes back to the present. She has not wanted to surf in two years, ever since her best friend died in the water. She looks like she's just seen the face of God.

I'm afraid it's the best sex she's ever had with me.

And it's all because I'm doing the 30 Day Experiment.

THE SECOND DAY

Linda calls and says she's in town. I haven't talked to her in two months. There must be some psychic signal I'm sending into the universe that says, "It's going to be really hard for me to have sex right now, so please come over and tempt me."

As soon as her lips touch mine, I'm hard. It is a different kind of hardness—urgent, independent, and definitely not going anywhere. She feels it and says, "I can always do that," as if she's responsible.

She says she doesn't want to have sex this afternoon, and that is fine. Just from the making out and rubbing, every nerve in my body is tensed and ready to explode. This gets more difficult each day.

I excuse myself to go to the bathroom, splash cold water on my face, and then return and tell her about the 30 Day Experiment.

That night, I talk to Kimberly on the phone. I'd messaged her on MySpace two weeks earlier. With her black bangs and large, innocent eyes, she reminded me of a Mark Ryden painting. She lives across the country in New York, but we've been talking nightly. She is easy to speak with, and the more I learn about her, the more I like her. Not only do we both collect 60's garage-rock and secretly enjoy being pushed around in grocery carts, but she is one of the sweetest, most genuine people I've ever never-met. Recently, I've been waking up thinking about her and randomly checking my phone throughout the day in case I miss a text from her.

I'd been wondering if she felt the same way about me. Tonight, I find out. After we hang up, she texts, "I'm rubbing my skin raw thinking about us. I hope you don't mind me admitting that to you."

I tell her that I don't mind and, six texts later, I know her favorite position, speed, and motion. While I'm having alphanumeric intercourse with Kimberly, Linda texts, "I want sex. Fuck your thirty days. Start it tomorrow."

Suddenly she's interested.

Then Kimberly texts, "My hips are moving so quick and high to meet my hand. I want to swallow you while I do it. Is that too much?"

Then Linda texts, "Baby, I want to fuck. Just one hour of bliss."

This kind of thing never happens.

Blood rushes to my pelvis. I feel like I'm going to pass out.

THE THIRD DAY

My friends think I've lost my mind. "Why put yourself through it?" they ask.

"Why does a man climb a mountain or walk on hot coals or read *Finnegans Wake*?" I answer.

I am doing it, first and foremost, to see if I can.

Rivers Cuomo, the singer and guitarist in Weezer, first planted the idea in my head. He was explaining that he'd recently taken a vow of celibacy as part of a Buddhist meditation program. This meant abstaining from not just sex but also masturbation. As a result, he said, he'd never felt more energized, creative, or focused in his life.

At the time, I interpreted it less as advice than as further confirmation of his peculiarities. But a few weeks later, Billy Corgan of the Smashing Pumpkins told me that he doesn't let his band have sex or orgasm on the day of a concert, so they can release all that power onstage.

Then, at dinner last week, I broached the topic, and a director at the table said that after he'd sworn off orgasms, he'd done the best work of his career.

As one of my editors used to tell me, it takes three to make an argument. So these three people, all far more successful than I am—combined with lingering adolescent self-flagellation guilt—inspired the 30 Day Experiment: No ejaculation for a month.

And today has been torture. Women I'm either sleeping with or want to sleep with have been calling nonstop. Then, worst of all, Kimberly decides to graduate from text sex to phone sex.

While we're talking about the Russian director Timur Bekmambetov, she starts breathing heavily into the phone.

"What are you doing right now?" I ask.

"I'm rubbing the outside of my panties." Her voice alone—candied, coy, and playful—turns me on. From the moment she said "hi," I was as hard as a crowbar—it doesn't take much these days. Now the pressure is too much to bear.

Rather than talking dirty to me, she just moans into the phone as she touches herself. This is actually hotter than ordinary phone sex because it seems more like we're doing it instead of just discussing it.

I bring myself dangerously close to the brink, then stop and take deep, calming breaths. I begin again, as she moans louder and sharper, breathes faster and shorter. I want her so badly. It feels as if there's a cord of sexual energy shooting from my body all the way to her in New York. I've never experienced anything like this during phone sex, probably because in the past I was too busy working toward my own orgasm.

After a few cycles of pleasure and denial, something else I've never experienced happens: my inner thighs and stomach—just above and below the crotch—begin tingling intensely. They feel simultaneously warm and cold, like they're covered with those icy-hot creams people use for pain relief.

"Did you come?" Kimberly asks after her orgasm subsides.

"I can't."

"What do you mean?" She sounds concerned.

I hesitate for a moment, then decide to risk explaining the 30 Day Experiment. There is silence on her end. She probably thinks I'm a freak.

"I want you to come," she pleads. "It makes me feel inadequate, like I wasn't good enough."

"You were so hot," I tell her. "I've never been that turned on over the phone."

She hangs up, dejected. I've tampered with the natural order of things. Women are so conditioned to expect a guy to come that when he doesn't, even if she has an orgasm, they tend to feel like the sex was incomplete.

I haven't met this girl yet and I'm already destroying her self-esteem.

Two hours later, my thighs and stomach feel like cold needles stuck into hot skin.

THE FOURTH DAY

Twelve times twelve is 144.

Eighteen times eighteen is 324.

Twenty-three times twenty-three is 529.

I can multiply any two numbers up to twenty-five together in an instant. I've become like a human calculator. It's an unintended benefit of the 30 Day Experiment.

Sex with Crystal isn't easy. After a while, even doing times tables in my head is no longer enough to hold back the tides of pleasure. I make her stop when she's on the brink of orgasm because I'm right there, too. She is not happy with this.

"Don't you enjoy orgasms?" she asks.

"I love to orgasm. It's like Nature's own heroin. That's why I want to see if I can kick it."

I now know how junkies feel. There is hardly a moment that goes by that you don't think about the rush. Every cell in your body screams for it. And the longer you have to go without it, the more consuming the desire for it becomes, until it drowns every other thought.

I suppose this is the other reason I'm doing the Experiment. I've been around some of the worst junkies in rock 'n' roll, yet I've never been addicted to anything: not even coffee or cigarettes. I used to tell myself this was because I didn't have an addictive personality.

On further reflection, however, I realized that I was addicted to one thing. Whether with a woman or alone, I'd had at least one orgasm a day for as long as I could remember.

To make matters worse, like most addicts, I've always been plagued by guilt about my habit. As a teenager, I used to think men were allotted only a few thousand ejaculations in a lifetime and I worried that I was using up my reserve too quickly. In college, every time I came, I thought it was somehow depleting my life force. And since then, whenever I masturbate, I feel not only dirty, but that it lessens my attractiveness and desirability when I interact with women over the course of the day.

The 30 Day Experiment, then, was not an option. It was a necessity. I needed to find out if I had the strength and willpower to break this addiction—and dispel the guilt-generated superstitions I'd been nurturing since puberty.

Of course, the Experiment would be much easier without all the sex, but by learning to enjoy the journey more than the destination, I'm becoming much better in bed. At least, I think I am.

"You suck." Crystal punches me playfully in the chest and dismounts. "I didn't get to finish."

"Maybe you're too orgasm dependent," I tell her.

Crystal is a six-foot-tall psychology student who used to pressure me to be her boyfriend. When I told her I didn't feel as strongly about her, she stopped sleeping with me for her own emotional health.

A month later, she changed her mind. "I decided you're too good not to share," she explained. The next week, I introduced her to Susanna and she had her first threesome. Since then, she's been willing to try anything once.

"I want to hear more about the orgasm thing and understand what you're trying to achieve," she says as I run to the refrigerator for water, enjoying yet another benefit of the 30 Day Experiment: no more rolling over and going to sleep. Sex now energizes rather than depletes.

I explain the rationale behind the Experiment to Crystal. She considers it for a moment, then asks, "Can women do this?"

THE FIFTH DAY

Kimberly is slowly taking the place of masturbation in my life. Every day, I look more forward to our bedtime conversations. Today, she confesses her feelings for me, and I'm not even scared. "I want to know you inside and out," she says. "I want to see a picture or a shirt or a toothbrush and know it's yours. I really, really care about you and what happens to you and how you feel."

I tell her that I have to speak at a seminar in New York in six days and am extending the trip to spend more time with her. We imagine every detail of our first night together, until she comes screaming my name. It is a sound that strikes me more profoundly than the greatest symphony or the most musical bird or the noise Windows makes when loading.

Afterward, I reach a new threshold of discomfort. The triangular area of flesh just above my dick feels tender and sore. And it is nearly impossible to take a shit, because when I squeeze my muscles, unbelievable bolts of pain shoot through the area above my crotch. When I look at the skin there, it seems swollen. But then again, I don't look at it that often, so maybe it's always like that.

It is now glaringly obvious that I'm doing this wrong. Something supposedly beneficial shouldn't hurt so much. In one of my favorite self-improvement books, *Mastering Your Hidden Self*, the author, Serge Kahili King, says that quitting a habit takes more than willpower. When you stop doing something, he explains, it leaves a subconscious void. And this void must be replaced by a new activity. This is why people who quit smoking cigarettes, for example, chew gum instead.

But I can't think of any type of gum—even Freshen Up—powerful enough to take away the urgency and pain I feel. The new habit would have to be something more physical, preferably an activity that alleviates the ache, like bathing my balls in cold sour cream.

I drift off to sleep, praying for a wet dream to relieve my burden. I've never had one before, probably on account of my compulsive masturbation. I'm awakened, however, by the phone.

"I want to do it with you." It's Crystal.

"Now?" I ask, horrified perhaps for the first time in my life by the prospect of a booty call.

"No, silly. I want to do the 30 Day Experiment."

I'm happy to have a female partner in restraint. I tell her about looking for a replacement habit and we decide on something constructive: exercise.

So, for the next twenty-five days, whenever I'm aroused, I'm going to do push-ups instead of masturbating. And I will master my hidden self.

THE SIXTH DAY

I'm getting turned on by everyone and everything. The words "polymorphously perverse" come to mind for the first time since college.

I spend twenty minutes scrolling through the numbers in my phone, thinking about women I've never even found attractive. I want to send them dirty texts and tell them to come over.

I hit the floor and do thirty push-ups. The blood begins to circulate through my body instead of amassing in just one place.

Later in the day, while I'm watching *South Park* on Comedy Central, an advertisement for *Girls Gone Wild* flashes across the screen. This is my first exposure to anything even resembling porn during the Experiment, and, in my weakened state, the montage of censored breasts and college girls making out seems like the greatest filmed entertainment our culture has ever produced.

I press the back button on the TiVo, and watch the commercial again, pausing to admire a few choice Mardi Gras revelers. As my hand slips under my belt, I have an epiphany: When I touch myself but don't ejaculate, I don't feel guilty or unclean. This means that I never had masturbation guilt; it was ejaculation guilt the whole time. And this makes sense. The trope that every sperm is sacred is hammered into childrens' heads, by everything from the Bible to Monty Python. Even in the second century, the philosopher Clement of Alexandria warned would-be auto-eroticists, "Because of its divine institution for the propagation of man, the seed is not to be vainly ejaculated, nor is it to be damaged, nor is it to be wasted."

So I'm not crazy: By wasting a load of sperm, I'm harming the future of my species. Or maybe I'm helping it. Depends on who you ask.

Thirty push-ups.

South Park is back on and I'm safe. The kids are on a road trip with Cartman's mother. And Cartman is calling his mom a slut and a whore.

I look at her, all crudely drawn circles and rectangles, and think that it would be awesome to sleep with her.

My hand is down my pants. I think I'm losing it: I'm getting turned on by Cartman's mom, or at least the demographic of desperate housewives that she represents.

Thirty push-ups. I'm going to be buff in no time.

Then Kimberly calls. She is drunk. She says she misses me. I miss her, too, and we've never even met. We have phone sex until every nerve in my body is tense and ready to explode. I start imagining what it would be like to pull out of her and just spray all over like a tube of toothpaste hit by a hammer.

I apologize for the simile. But I keep teasing my body and it's taking its revenge on my mind.

More push-ups. Until I can't do any more.

I can't go on like this.

Perhaps it's not enough to simply swap habits. The entire concept of the Experiment could be a misunderstanding of the wisdom of Rivers Cuomo. Maybe the magic energy shift happens not through refraining from shooting out a milky white fluid, but from actually being desireless. This is, after all, what most great spiritual disciplines advise. To paraphrase the Buddha, craving leads to suffering. And I am definitely suffering, which is pathetic considering that it's only been six days.

THE SEVENTH DAY

Crystal calls and updates me on her first day of self-restraint. Unlike me, she did due diligence. With Google on her side, she discovered a spiritual backbone to the Experiment that I've completely neglected—more out of laziness than ignorance.

"You're just withholding and that's not healthy," she says.

"I know. It hurts when I sit now. I'm worried that I'm going to get prostate cancer or something."

"See," she says self-righteously. "You're supposed to take the life energy and, instead of holding it back like a dam, circulate it through your body."

"And how does that work exactly?"

"It's supposed to be done with a partner," she hints.

She sends me links to Taoist and Tantric Websites with information about sexual gurus like Mantak Chia, Stephen Chang, and Alice Bunker Stockham. From Stockham's research, I learn a new phrase: "coitus reservatus"—sex without ejaculation. From Mantak Chia, I learn that it's possible to have an orgasm without actually ejaculating. And from Stephen Chang, I learn the deer exercise, which is based on ancient Taoist monks' observations of the potent, long-living deer, specifically the way it wiggles its tail to exercise its butt muscles. The ritual is supposed to spread semen from the prostate to other parts of the body. I need to do this immediately.

I sit on the toilet with my laptop computer open at my feet and follow the directions, rubbing my hands together to generate heat, then cupping my balls. I place my other hand just below my navel and move it in slow circles. Then I switch positions and repeat. For some reason, I can't imagine a deer doing this.

For part two of the exercise, I tighten my butt muscles, imagining air being drawn up my rectum, and hold it. Then I relax and repeat. It is sort of like doing butt push-ups.

The pain persists, but now it's mingled with embarrassment. I'd rather get caught masturbating than doing butt push-ups.

Before going to sleep, I call Kimberly and attempt Mantak Chia's method of orgasm without ejaculation, hoping it will provide some relief.

When she pulls a dildo out of her bedside table and narrates its next moves in graphic detail, I can't take it anymore. I press on my perineum, tighten my

PC muscle, and do a butt push-up. It just barely holds back the flood. However, I don't have a dry orgasm, either.

"Oh my God, I just came so hard, baby," Kimberly gasps. "Did you come?"

"I can't yet." All I've done is make the pain worse. Why do I keep doing this to myself?

There is silence on the other end. It is not a comfortable silence.

"I'll tell you what," I decide. "When I see you in New York in four days, I'll really come. I think it would be amazing to end this experiment with you."

"But what about the thirty days?" she asks, more relieved than concerned.

Fuck the thirty days. I am willing to fail this experiment for what may be love. In fact, any excuse to end it will suffice.

THE EIGHTH DAY

As I attempt another of Crystal's ridiculous exercises—the straw meditation, which involves imagining the orgasmic energy being sucked up my spine and into my head—I remember the night I learned to masturbate.

I was at overnight camp in Wisconsin, and for some reason I will never comprehend, the two cool kids in my cabin decided to show everyone how to beat off.

I lay in my top bunk bed and watched as Alan snuck into the counselor's area and returned with a red can of Gillette shaving foam. He stood in the center of the room in his blue camp shirt and dirty white briefs, as if performing theater in the round, and addressed the nine other pubescent boys of the Axeman 2 cabin.

"Just squirt some into your palm. Then you gotta move your hand like this." He stuffed his fist into his underpants and began the demonstration. His loyal follower, Matt, hopped off his bunk bed, squished out some shaving cream, and joined him.

We were too young to know that masturbation was supposed to be a private act, its revelation to peers punishable by mockery and ostracism. In my presexual brain, it was just another group activity, like archery or orienteering.

Hank, sickly and effeminate, rolled out of bed and distributed dollops of shaving cream to everyone else in the cabin. We all got to work.

The sight, in retrospect, was ridiculous. People often wish to be innocent again, but there is no such thing as innocence. Only ignorance. And the ignorant are not blissful; they are the butt of a joke they're not even aware of.

I didn't come, or even feel much pleasure. I don't remember if anyone else came, either, but, according to Alan, that was the goal. It was a race and, after camp ended, Hank won: He wrote me a letter, excited because he'd masturbated and "a few drops of come even came out."

Almost a year later, lying in bed at home, I began pulling at myself one night. I thought of a story a friend had told about going to the movies with a girl from school and getting a hand job. I extracted every detail from him: I'd never kissed a girl before, or even been within kissing range.

As I touched myself that night, I imagined it was me getting that hand job in the movie theater. Soon, pressure begin to build and I seemed to be separating from reality. My breath caught in my throat, my body was seized by what felt like rigor mortis, and then it happened. A small pool dribbled out of the tip. I reached over my head and turned on the reading lamp next to my bed, careful not to mess it up. Then I conducted an examination. Because of the way Hank had described his come, I thought it would be clear, like raindrops. But instead it was a little viscous puddle with swirls of cloudy white and a few transparent patches.

As I write this, I realize for the first time why my sexual fantasy is fooling around in public places like clubs, theaters, and parties, where no one can see what's going on, since that's the image to which I had my first orgasm.

"You have to check this out," I told my nine-year-old brother the next day. "Follow me."

He padded into the bathroom behind me. I stood on his toilet, dropped my shorts, and thrust my hips out so that when I came, it would dribble into the sink and not make a mess. Then I got to work.

Outside of sweat and tears, I'd never known my body to make a product that wasn't waste. I was proud. I was an adult now.

THE NINTH DAY

I wake up next to Gina. She'd stopped by after bartending the night before for a quickie. But it was 3:00 a.m., and in addition to being tired, I was desireless. She took it personally.

"You're over this, aren't you?" she asks in the morning.

"What do you mean?" I protest, though I know full well what she means. In addition to my new effort to limit my desire, ever since I'd started talking to Kimberly, I'd grown more distant. "Is this because I didn't have sex with you? In twenty-one days, everything will be back to normal."

"It's not that. I love you, but I have to love myself enough to realize that you don't want this."

Above my bed, there's a small painting she made for me in happier times. She takes it off the wall and lays it in her lap. I watch her, sitting upright in the bed, her hands shaking as she struggles to remove the backing on the frame. The latches holding it in are too small and stubborn for her trembling hand.

She eventually clicks them open and pulls off the back of the frame. Instead of removing the painting, she takes the backing, pinches the black paper on the inside, and tears it off. Beneath, there is a hidden note she'd evidently written when she first gave me the present. I never even knew it was there.

She throws the torn backing onto my chest, then walks out of the house. I pick it up and read:

"You will be a great husband one day when you are ready and find the one. You will be an amazing father to cute intelligent baby Neils. You are going to hurt me. But I will always love you."

My face begins to swell, my eyes and nose feel warm and flushed, and suddenly tears begin leaking out.

I'm going to miss her. And I will always respect her: the picture frame gambit was the work of a true breakup artist.

THE TENTH DAY

Tomorrow, I'm finally going to see Kimberly. As my other relationships have fallen apart, she has remained loyal. I feel like we've met before, slept together before, pushed each other around in grocery carts before. There are moments when I actually think I love her, but I know it's just a combination of attraction, obsession, and curiosity. I'm sure she feels the same way about me.

That is, until she calls to tell me she has to take a last-minute job as a production assistant in Miami and won't be able to meet in New York.

"I don't have a choice," she says. There is a hostile, self-defensive tone to her voice that I've never heard before. "I really need the money. I have like thirteen dollars in the bank right now."

I'm crushed. I've been so fixated on meeting Kimberly in New York that I can't imagine being there without her. I start to tell her that.

"Don't," she snaps. "There's nothing I can do."

"I'm not upset," I say, upset. "This is just really unexpected. But it's not the end of the world. Maybe I can visit you in Miami after New York."

"I may have to disappear for a few days," she says, her anger melting into tears. "I just need to think about us."

The more we talk, the more emotional she gets. The more emotional she gets, the more she withdraws. "So you're not going to meet me in New York and you can't make a plan for Miami?" I feel like she's put a cigarette out in my heart. "I need to know that I'm going to see you."

"You're making me cry." She's yelling at me now. I'm dealing with emotions; my logic is useless, my anger counterproductive. All that's left is frustration, paranoia, and a sickness in every cell in my body that was anticipating the end of the 30 Day Experiment tomorrow and the beginning of a fairy-tale romance.

"If you have to disappear," I press, "then first give me a time when I can see you, so I have something to look forward to. Otherwise, this has all just been a fantasy relationship."

"A fantasy relationship?" Evidently, I've said the wrong thing again. "I wanted to see you so badly and you know that. I wanted to be your girlfriend." She stops sobbing, then hits me where I'm weakest. "Don't blame this on me. You're the one who's impotent on the phone."

On a more positive note, after we hang up and I collapse onto the floor of my bedroom, I realize something: My balls haven't ached all day. I seem to have made it through the pain period.

THE ELEVENTH DAY

The next afternoon, I'm in a cab to LAX to take a plane to New York. At the same time, Kimberly is in a cab to JFK to take a plane to Miami. Neither of us has slept. We spent the night arguing, showing each other our worst sides. And now we are texting each other the ugliest good-bye in the world: "Have a nice life."

On the plane, I'm a wreck. Sleepless, unshaven, blanched, I hold my head in my hands the whole ride and replay the conversation in my mind, regretting all the stupid things I said and wondering if Kimberly sabotaged the relationship on purpose. Perhaps she's scared to meet, worried that either she'll disappoint me or I'll let her down. Perhaps she never planned to meet in the first place because she has a boyfriend in Miami or is a lunatic telephone stalker or has a fake MySpace profile and actually looks like a linebacker.

None of these possibilities alleviates the heartbreak. I didn't know I could feel this way about someone I've never met.

The empty bed fills my hotel room like an accusation. I'd spent so many nights imagining lying here with Kimberly, seeing each other naked for the first time, acting out all our phone fantasies, taking a candlelit bath together, and then getting under the covers and talking until we fell asleep in each other's arms. I feel like a fool for trusting her, falling for her, spending all those hours on the phone building a future with her that she knew would never exist. At the same time, I wonder how much my infatuation with her was a result of transference from the 30 Day Experiment: replacing one addiction with another.

I decide to go to her favorite lounge in the city, Amalia, to search for someone just like her. Instead, I find Lucy, a young, thick Brazilian girl with a lisp, a too-tight black dress, and no interest in 60's garage rock or grocery carts.

She follows me around Amalia, touching me at every opportunity. So I tell her, not really caring whether she accepts or rejects me, "We should take one of these girls home with us tonight."

It is presumptuous and I prepare for her to snap back, "Who says I'm going home with you?"

But instead, she snaps back, "We should take, like, five of them home."

"Who's your favorite?"

She points to a tall, frail girl with pale skin, long auburn hair, and a big, toothy smile.

Two hours later, my hotel bed is full. Lucy takes my computer and plays a Shakira video online. Then she rises and lisps along in perfect harmony while working her hips in slow circles. The tall girl, an off-Broadway actress named Mary, lies in bed on her stomach and watches. By the end of the dance, she's on her back and we're making out.

She gets the chills every time I kiss and bite her neck, each shiver shaking off a little more inhibition, until she tells me, "I want to see your cock."

I'm taken aback by her sudden boldness. It seems less like she's turned on and more like she's decided to play a role.

"Get naked," she orders. "I want to see it."

I play along, and within seconds I'm completely nude. They're both still wearing their dresses. Without clothes or even actual desire, I feel awkward. I miss Kimberly.

"I want to watch you fuck Lucy's tits."

Having something to do helps. Lucy joins us on the bed and removes her shirt. I kneel over her, put my dick between her breasts, squeeze them around me, and start sliding up and down. It is as unsexy as it sounds.

"I like watching you fuck her tits. I want to see you come all over her."

On that command, I lose what little arousal I was able to muster.

"There's something I should tell you," I begin.

They both tense, assuming the worst.

"No, it's not that."

After I explain the 30 Day Experiment, we start fooling around again. But it's not the same. Mary eventually gathers her clothes and leaves, and Lucy falls asleep while I'm going down on her.

It is the worst threesome ever and I don't care. I am beyond desire. But I am not beyond loneliness.

When I reach over to the nightstand to check my phone, I notice a text message from Kimberly. My heart clenches. I feel excitement, anxiety, curiosity, fear, and, when I see the message—"Are you phonable?"—relief.

Careful not to wake Lucy, who's lying naked and spread-eagled over the sheets, I slip into jeans and a T-shirt and tiptoe into the hallway. There's a window ledge next to the bank of elevators, and I perch there and call Kimberly.

"Hey," she says. I adore her voice. It is the sound of gravity sucking me into her world. I never thought I'd hear it again.

"I'm glad you texted." I want to tell her that I wish she were here, but I know it will upset her. "I'm sorry for overreacting. I just had my heart set on seeing you."

"I did, too. I really thought we would be together, like, really be together. But last night changed things. I saw another side of you."

"Yeah, I understand. I think the relationship went as far as it could go on the phone, until there was nowhere to go but down."

We spend the next hour trying to talk things back to the way they used to be. Eventually, we succeed. "I wish I could be with you right now," she whispers.

Minutes later, I'm squeezing myself through my jeans. "I'm imagining you fucking my face," she is saying. "You're just grabing my head and thrusting into my mouth, as hard as you can. And you're reaching down my back and putting a finger inside."

I'm not sure if this is even physically possible, but it's making me feel like I'm thirteen again and stealing my father's copies of *Penthouse* to read the letters. I undo the button of my jeans and reach into my pants.

I imagine the night as it should have happened. She is here, in my hotel room, pale body against the crumpled sheets, lips swollen and chin red from endless kisses, thighs wet from . . .

I hear an elevator whirring, people laughing. I don't stop. I'm half-exposed. The pressure is building, the body is separating. *Wet from* . . . This is the night I was supposed to end it all, the night of the toothpaste and the hammer. *Thighs wet from* . . .

I lower myself into her. I could stop. I should stop. I can't stop. She's coming, I'm coming.

I watch it release. It doesn't fly everywhere the way I expected and, on some level, hoped. It just flows out, into a giant pool, like the first time I ever came—except this time, instead of fantasizing about a public place, I'm actually in one.

An immense wave of relief spreads through every nerve ending, my eyes fill with tears of joy, and white fireworks explode lightly in my head.

"Did you come?" she asks.

"Yes." I already feel guilty: less for masturbating than for not even making it halfway through the 30 Day Experiment.

"I can't believe it took me so long to get you to do that." She pauses and I hear her suck in air. She's having an after-phone-sex cigarette. "You were giving me a complex. I thought: I'm no good. I'm not turning this man on, and he's giving me all these orgasms."

I suppose she needed the closure. And so did I. We basically had an entire relationship over the phone: we met, fell for each other, dated, had sex, fought, and broke up without even meeting. Tonight was just makeup sex.

It is clear that we will never meet. Like the idea that I could actually go thirty days without an orgasm, the relationship was just a pipe dream.

Before I go to sleep, I call Crystal in Los Angeles. She's handling the experiment just fine: no pain, no anxiety, no attraction to cartoon characters. But she's of a different gender, the one more likely to hurt after the orgasm than before.

I tell her about the benefits of the Experiment: I've been less tired during the day, possibly attracted more women, and definitely saved on Kleenex. Then I tell her about the downside: I failed. As she tries to console me, I realize that I actually set myself up to fail. I went on a diet, then hung out at Baskin Robbins every day.

The Buddhists are right. Desire is my pilot. Most of each day is spent giving in to it. When I'm not fucking, I'm chasing. When I'm not chasing, I'm fantasizing. I have had sex with tens of thousands of women in my mind. And now that the Experiment is over, they will be back. All of them. A parade of innocents. The college girl swinging her hips through the supermarket aisles. The secretary posing at the crosswalk as I drive past. The party girl making out in the hot tub on the reality show. The girls who have gone wild. Cartman's mom. Kimberly. If I can't have them in real life, I will have them in my imagination.

I am an addict.

I am a man.

RULE 11
NO MAN WINS THE GAME ALONE

I.

Love is a velvet prison.

That's what I think when Dana rolls on top of me. Her eyes are shining, her lips smiling but not too much. She doesn't have to say it, but she does.

"I love you."

And then I feel the bars come down around me. They are only made of velvet. I have the physical strength to escape, but I don't have the emotional strength. And so this velvet is thicker than iron. At least I can bang my head against iron.

She looks at me, expectant, awaiting a reply. I can't speak it. I'm doing all I can to keep my eyes open. I want to go to sleep. I want her off me. Her emotions are now my burden. The wrong look, word, or gesture can singe her like a poker.

She lies on top of me, naked, her eyes searching for something in mine. When she doesn't find love, she will settle for hope. And so I am trapped. In this velvet prison.

II.

"If one of your skanky fucking whores calls and hangs up on me again," Jill fumed, "I will kill her."

"What are you talking about?" I never knew what sort of mood she'd be in when I walked through the door. "Who did what now?"

"One of your whores called," she yelled. "She said it was the wrong number, then she hung up."

"Did you ever stop to consider that maybe it actually was a wrong number?"

"Oh, she knew," she spat. "She knew it was me. The bitch."

I left the house, climbed into my car, and drove down the Pacific Coast Highway. I'd seen Jill work herself into such a frenzy over the skanks and whores I'd slept with in the past that her mouth would actually foam. I had to get my life back.

I used to tell girls that if relationships were a funnel, I wanted a woman who would travel with me up to the wide side. I never realized the inaccuracy of the metaphor until that drive: Funnels only go one way, toward the narrow side.

III.

You can smell Roger a block away. He sleeps in the streets of Boston and yells at lampposts. The people at a local bookstore who look after him tell me he was drafted to play major league baseball in the early seventies. One day, though, someone slipped acid in his beer as a joke. He was never the same.

Roky had a small, influential rock 'n' roll hit in the late sixties. Arrested for possession of a joint, he pled insanity to avoid a jail term. Successful, he was sent to a sanitarium, where years of electroshock and Thorazine treatments melted his mind. In 1981, he signed an affidavit stating that a Martian was in full possession of his body. At age fifty-four, a mental and physical wreck, he was put into legal custody of his younger brother.

My grandmother had a stroke when she was in her seventies. Afterward, she regressed to the age of thirty-two. She no longer recognized my brother or myself, and instead spent every day waiting by the telephone for her mother to call from the hospital. Her mother had died in the hospital forty years before.

There is just a thin string connecting each of us to reality. And my biggest fear is that one day it will snap, and I'll end up like Roger or Roky or my grandmother.

Except, unlike them, there will be no one to take care of me.

POSTFACE

"Kind of a cynical ending, don't you think?"

"I wouldn't say cynical. Maybe sad. Or afraid."

"After the way you've carried on with all these women, do you expect me to feel sorry for you or something?"

"That's the last thing I'd expect, especially from you." In the years that had passed, the scene hadn't changed. The producer, his houseboy, his dog didn't even appear to have aged. He was a creature of habit. And one of those habits was pointing out the inconsistencies in my thinking.

"So it's just about you feeling sorry for yourself then?"

"It's more about feeling confused. I wrote the stories you just read after the failure of two relationships. Afterward, I talked to hundreds of married men and women who felt unhappy or stuck. And I just want to make the right decisions in life."

"I see." The book manuscript sat on top of a blanket in his lap like an offensive drawing made by a schoolboy. "So why did your last relationships fail?"

"I guess they failed because the women developed certain behaviors that made me doubt the success of a forever-type relationship with them."

"And I suppose you didn't have anything to do with the development of these behaviors?"

I had walked him right to his moral high ground again. "Of course I did. It always takes two."

"And now you've decided to be alone and miserable forever?"

"I just tried so hard to make these relationships work."

"How exactly did you try?"

"I was honest. I was faithful. I cut off all the other women I was seeing. I

didn't tell lies or carry on secret flirtations or sneak around behind their backs."

"And that's how you make a relationship succeed? By not having affairs with other women? That's like saying you learn to swim by getting in the water. It's a given." The sun began to sink beneath the ocean outside his picture window. "Did you ever stop to consider that you never really tried?"

"What do you mean?" The houseboy set a ceramic bowl of cherries in front of him, then lit a stick of nag champa incense. I was walking right into some sort of trap of his.

"You worked really hard to learn the game. You read every book there was, traveled around the world, met all the experts, and spent years making countless approaches to perfect the craft."

"I think I see what you're getting at."

"And what do you think that is?"

"That maybe I need to learn how to have a relationship in the same way I learned the game."

He slowly, triumphantly plucked a cherry off its stem. "Ultimately, you're going to have to make a choice at some point in your life. And that choice is to decide: Do you want to find a woman to spend your life with and make a family together? Or do you want to keep giving in to your impulses and continuing to have sexual adventures and relationships of varying lengths until you can't anymore?"

"Doesn't sound like much of a choice."

He popped the cherry into his mouth and sat contentedly on the sofa. I used to think that his slow gestures and exaggeratedly calm demeanor were an affectation, a sign of faux spirituality. But I'd since come to envy his stillness of mind.

"So let's say I choose to be with someone forever," I continued. "You're saying that I need to make that relationship a project and devote the energy I once used chasing women to getting better at it."

"Yes."

"Yes and?" He was holding out on me.

"And the challenge is to find someone to love who not only loves you in return but is also willing to work with you on this life project."

"That's easier said than done. How do you know when you've found the right person?"

"When you're with someone you grow closer to over time instead of apart from," he said. "A lot of people make the mistake of trying to defend principles in relationships. My goal is long-term happiness. And I make choices that aren't going to undermine that goal. Even if it means giving up a freedom in exchange."

"Man, that's scary." I hated that he was winning. I hated that the answer had the word *work* in it. I hated the idea of making a decision that closed other doors of possibility and experience behind it.

"Or exciting. As with learning anything, it will be difficult and there will be obstacles, but eventually you'll master it. And you'll find a strength and confidence that no amount of one-night stands and threesomes can ever give you."

"That all may be true, but there's still one problem we haven't solved." He listened intently. Solving problems was his specialty. "So what happens a few years into the relationship if I feel the call of the wild and just want to go have sex with someone new? How do I control that, or not resent her for keeping me from those experiences?"

"Well," he said patiently, "you think about how that would affect the project you've dedicated your life to. People who work in banks generally don't steal the cash. Although they want more money in the moment, they value their future more."

In the intervening years, I had interrogated many men in long-term relationships. Most of them simply gave in to the call of the wild and slept with other women behind their partner's back. But that is a recipe for disaster. Even if she never finds out, the guilt, secrecy, lying, and sneaking around eventually destroy the love a couple once had. An honest alternative is an open relationship. However, the couples I met in open relationships not only still had drama, but were no longer in love. They were just codependent.

But there are other options. "I suppose if I still wanted to have my cake and eat it, I could explore swinging or polyamory or being with a bisexual girl."

"If she's comfortable with that, I suppose it's something you could try." He paused and stroked his chin. I saw a glint in his eyes. "But there's something you need to know first."

"What's that?"

"When I was reading over that discussion of ours, I realized something." He took a sip of water. I knew that it wasn't due to thirst, but a sense of confidence that his next words would reveal my complete idiocy. "That whole idea of not having your cake and eating it—the expression is wrong. The saying should be: you can't eat your cake and have it."

"I'm not sure I get the meaning."

"It means you should be glad you were lucky enough to experience the luxury of a cake in the first place. So stop staring at it and worrying about what you'll lose by committing to it—and start enjoying it. Cakes were meant to be eaten, not collected."

I hated him sometimes. For being right.